MASTERING NUMBERS

Helping to develop Confidence, Accuracy & Speed

Videos -:- PowerPoints

Self-correcting worksheets are available

Reliance on a calculator
Does NOT make you smart
It is You who does

Chris O'Brien

ModMaths Pty Ltd

Copyright © 2013 Christopher John O'Brien

ModMaths Pty Ltd
Email   mastering_numbers@modmaths.com.au
Skype   modmaths

Publisher: Amazon Createspace

National Library of Australia Cataloguing-in-Publication data
Author: Chris O'Brien
Title: Mastering Numbers : helping to improve confidence, accuracy & speed
ISBN: **9780987488404 (pbk.)**
Subjects: Mathematics—Study and teaching.
Dewey Number: 510.71

Reproduction and communication for educational purposes
The Australian Copyright ACT 1968 (the Act) allows a maximum of one chapter or 10% of this book, whichever is the greater, to be reproduced and communicated by any educational institution for its educational purposes provided that the educational institution (or the body that administers it) has given a remuneration notice to Copyright Agency Limited (CAL) under the Act.

For details of the CAL licence for educational contact:

Copyright Agency Limited
Level 15, 233 Castlereagh Street
Sydney NSW 2000
Telephone: (02) 9394 7600
Facsimile: (02) 9394 7601
Email: info@copyright.com.au

Reproduction and communication for other purposes
Except as permitted under the Act (for example, any fair dealing for the purposes of study, research, criticism or review), no part of this book may be reproduced, stored in a retrieval system, or transmitted in any form or by any means without prior written permission from ModMaths Pty Ltd. Enquiries concerning reproduction outside the scope of the above should be made to the author at the email address above.

Every effort has been undertaken, in the preparation of this book, to make sure that all calculations and answers are correct. However, there may be some errors or omissions. The author would welcome any correspondence to enable correction or additions in subsequent editions.

Notice of liability
The information contained in this book is distributed without warranty. While precautions have been taken in the preparation of this material, neither the Author nor ModMaths Pty Ltd shall have any liability to any person or entity with respect to any liability, loss or damage caused or alleged to be caused directly or indirectly by the instructions and content contained in the book.

All efforts have been made to gain permission for the copyright material reproduced in this book, but the publisher may not have been successful in contacting all the copyright holders. The publisher welcomes any information that will enable rectification of any reference or credit in subsequent editions.

Front & Back Cover
Designed by Scarlett Rugers,
www.scarlettrugers.com

Cartoons
Created by Randy Glasbergen,
www.glasbergen,com

Table of Contents

	Introduction	5
Chapter 1	Addition	11
Chapter 2	Ahhhh subtraction	19
Chapter 3	The art of multiplication	32
Chapter 4	2 digit by 2 digit multiplication	70
Chapter 5	Shortcuts in Multiplication	77
Chapter 6	3 digit by 3 digit multiplication	123
Chapter 7	Digit sum: Checking by 9s	145
Chapter 8	Division	152
Chapter 9	The Trachtenberg method	174
	Conclusion	187
	Index	188

I dedicate this book to my children: Joseph, Samuel, Dominic and Annie; who have survived having a father who plays with numbers.

I express my love and gratitude to Leanne: for her patience, support, belief but most importantly, for her unconditional love.

Introduction

A teacher walks into a combined Year 3 – 4 class, holding three chocolate frogs. He asks if there are any students who like reading stories. More than a few hands shoot up. Choosing two students, he invites them to recite the alphabet out loud. With great confidence they begin ABC… often in the *Twinkle Twinkle* version. (Since this teacher lived in Australia, if a student ended with Zed, not ZEEEEEEE, they received a frog) Selecting a third student, he takes out a large block of chocolate, and asks them if they are ready?

"Good, please say the alphabet backwards."

Many stutter, catch their breath, begin Z, Y … X …………W ……………………….. and then they fade into silence.

Once he has their attention, he writes the following multiplication on the board.

$$342,618,032,541,724,351 \times 11$$

"Within 20 minutes, you can learn how to solve this multiplication. The students often respond by saying, "No Way!"

However, you do not have to be a *Maths genius* to solve this, you just have to know how numbers work.

'*SECRET*' information for Parents & Teachers

Why are people able to say the ABC … Z fluently, yet many have difficulty reciting the 7 times table? Right from a very early age, parents and teachers have 'sung' the ABC …. to their children, who now can say it without thinking of the next letter. But as soon as they are asked to say it in reverse, many struggle to say it fluently.

I came to teaching 12 years after leaving school. Choosing to teach Mathematics was a natural fit, as numbers make sense to me. Finding patterns and enjoying the challenge of finding the solution to a problem interested me. When people asked me what I do for a living and I replied that "I teach Mathematics," often their response was, "I was no good at Maths", or "I hated Maths."

It is rare for English teachers to receive the same response, "I was no good at English," when asked what they do for a living.

From our earliest school experience, sitting in a Maths class, we might remember the tension that happens when the teacher is about to ask a question. As the eyes of the teacher scan the class ready to select the 'victim', many students have their heads down feeling some dread, sensing though that there are a few hands shooting into the air bursting with confidence. This time it's Joe's or Jessica's turn. A collective sigh is emitted.

When some students are asked a Maths question, they can experience a burning flush rising up to their cheeks, they sense all the eyes shifting to gaze in their direction, feeling a sense of dread as the teacher and class wait for the answer.

Are they right? Yeh! Phew! Or are they wrong? Oh No! It might be me next.

What students soon learnt was that there are some kids who get numbers and there are others who do not. But what is evident is that students, *who don't*, still carry this dread into high school and beyond. Over many years of teaching, I would see senior students reach for their calculator to answer the simplest of mental calculations.

Before we journey into the world of Numbers there are two stories I would like to share.

During my first year of teaching, students were asked to estimate the number of bricks needed to construct a standard 140 m^2 house. For the next two weeks, the students learnt all about measurement, area and estimation. Confident they would do well, their responses were collected. Opening the first student's paper, there on the page in bold print was

To construct a 140 m^2 house you need   51 020 758 bricks.

This beautiful number, written down accurately from the calculator screen, is just a *'little bit'* high. The look on the student's face was priceless, "How could I be wrong? That is the answer that the calculator gave me!"

The second story comes from Isaac Asimov's book, *Nine Tomorrows*. It has nine short stories on different possibilities of the future. In one story, set in the sixty-fifth century, teenagers in their eighteenth year were assigned their careers after their brains were scanned and then imparted with all the knowledge that their selected career needed. Those who were the 'best' candidates were sent to the plum careers in the outer worlds. George Platen dutifully underwent this scan and the result came back, REJECTED, FEEBLE MINDED.

Everyone was mortified. George was so popular. He was going to be a computer programmer. But the scan said that he was not suitable for any job and the scan never lies. The family was distraught and some of his friends began to drop away. All were asking how could George be a failure? George on the other hand, after the initial shock, refused to accept the validity of the scan. He was not worthless, he was not a failure. Being classified as feeble minded, George is exiled to an institute that studies these 'unfortunate' people. George refuses to accept his classification. He fights the nameless system that pronounced his fate. The upshot of the story was: **Who are the people who can push forth the boundaries of knowledge?** For these people have to have the capacity to think, to search, to break through perceived limitations of existing knowledge. George was one of the few who had that ability, but first he had to accept that he was more than what other people 'boxed' him to be.

When some students meet a new mathematical skill they often respond, "When will we ever use this after school?" Off course, there are some technical things where we do not need to know why or how they work. However, everyone needs to know how to think, to learn and apply new skills, to examine patterns and relationship and to solve problems. This is the challenge of Mathematics.

There are many occupations where people do need to make sense of numbers such as nurses with medications, engineers with calculations and trades with measurements and estimations. Many businesses fail on account of the owner's inability to work with numbers and gain an understanding of the world of Finance. Mathematics also helps develop the processes and skills required to play games, to solve problems and apply strategies.

Today's 'I' generation, the world of the internet, the iPhone and the iPad, are people who have a vast array of information at their fingertips. There are many dynamic learning tools and fantastic apps available on iTunes or Android systems that can help people learn. In addition, calculators and computers are ever present to do the calculations for us so that we do not have to think. However, it is my firm belief that **"reliance on a calculator does not make you smart. It is You who does."**

For instance, {without using a calculator} does $4 + 2 \times 5$ equal 30 or 14?

On most non-scientific calculators (often the ones used in business) the answer on the screen is 30. The actual answer is 14.

The scientific calculators used in school try to mirror the correct method of applying Maths to solve problems. But try the following using a scientific calculator -6^2. Many will show the number -36. However, the correct answer is 36.

If students rely on the % button to calculate the percentages, what happens if the calculator used at work does not have a % key? The following example provides an example.

A store offers a general sales discount of 10%. They also offer an additional discount of 5% if you pay by cash.

If an item has a price tag of $100, and you pay by cash, how much will the item cost?

At first glance the logical answer seems to be $85,

The correct answer is $ 85.50.

How can a book like this, which by its very nature is a static learning device, help to improve your child's and your ability to master numbers and help develop mathematical skills? This book is designed to help you understand the processes of working with numbers, to look at patterns and skills that provide an alternative way of performing mathematical operations, and giving you confidence to rely on something that never has a power failure or flat batteries- your own mental abilities.

Available through the author's website modmaths.com.au are PowerPoints that take you step by step through each calculation and example in the book. You can download many self-correcting Excel worksheets of the techniques explained in the book.

PREMISES

There are a number of premises underpinning this book.

The first is: Calculators do not make you smart! If you do not understand how to manipulate numbers then if you only write what the calculator's display gives you, then you will not be able to judge whether the answer is correct or not.

The second is: A computer ONLY based Mathematical software programs that requires the answer to be entered in on the screen, does not necessarily provide the skills to manipulate numbers when you are away from the computer. These programs and app game based strategies are fun and can help a person develop their mathematical ability. However, if a program concentrates on achieving an answer rather than the process to get the answer, your ability to interpret the answer on the display might not be developed. A child, teenager or adult needs to know that an average size house cannot use 51 million bricks in its construction.

Computers produce data that has to be turned into information. Data needs to be interpreted to be analysed, to be made sense of and then communicated as information from which people can make meaningful conclusions and can make meaningful conclusions and predictions. Although we have a greater reliance on technology, people still need to have numbers making sense.

The third and most important premise is: Making sense of numbers requires hard work and dedication.

A number of years ago at a conference on Improving the Teaching of Mathematics in Schools, a well-known neurologist shared the following story. Two groups of twenty mice were injected with a disease which affects the brain. The first group was kept in a large flat container where they were given plenty of food and drink but nothing to play with. The second group were kept in the same size container, with the same quantity of food and water, but had lots of activities and things mice like to play with. After 3 months the mice were killed and their brains examined. Over 90% of the mice in the first container had severe onset of the disease, while less than 10% of the mice from the second container developed the same degree of degeneration, with many showing little or no signs of the disease.

Activity, practise and play, helps keep our brain healthy. By doing these things we develop and reinforce the neural connections that enable us to complete skills and activities. If you have ever watched a child or skilled adult play a game, they can do it quickly and skilfully because their neural paths and connections have been established over the hours of practise and repetition.

NO MAGIC WAND

There is no magic wand that will suddenly enable you or your child to do all mathematical sums in the blink of an eye. However, you can learn skills and techniques that may help your confidence improve and lessen any 'fear' you have when working with numbers. To learn to say the alphabet backwards in less than 10 seconds, will require practise. But if you keep trying, one day you will be able to say it without thinking of the next letter. This is true for the mathematical techniques that are covered in this book.

The underlying skill that will be carried through all calculations is pattern recognition. The focus of the first two chapters of this book is to examine the skills and abilities in addition and subtraction and show strategies that might improve your accuracy and speed. Chapter 3 uses the Trachtenberg system to show you how to apply patterns and speed strategies in the mental calculations up to the 9 times table.

The techniques to solve harder problems are made easier if you are able to do a 2 to 9 times tables in the same speed as it takes to click your fingers. Chapters 4 to 6 concentrates on demonstrating the techniques to do 2 digit by 2 digit multiplication and 3 digit by 3 digit multiplications using the Vedic pattern and other 'tricks" that will help to solve these calculations very quickly compared to the way you might have been taught at school. Chapter 7 shows the system of 9s as a checking without the use of a calculator. Chapter 8 examines a method to do division that requires only 2 lines of working out. Chapter 9 explores the Trachtenberg system of units and tens to solve multiplications which can be extended beyond the 3 digit calculations. An Adobe PDF of the algebra underpinning these techniques can be downloaded from modmaths.com.au.

Chapter 1

"You have to solve this problem by yourself. You can't call tech support."

Addition

Many children learn in primary school that each number from 1 to 9 has a 'best friend'.

$$1 \Leftrightarrow 9$$
$$2 \Leftrightarrow 8$$
$$3 \Leftrightarrow 7$$
$$4 \Leftrightarrow 6$$
$$5 \Leftrightarrow 5$$

Each pair of numbers sum to 10.

Find the missing value of the following sums.

1 + 9 = ___	2 + ___ = 10
3 + 7 = ___	4 + ___ = 10
4 + 6 = ___	___ + 3 = 10
8 + 2 = ___	1 + ___ = 10
5 + 5 = ___	5 + ___ = 10

When two numbers have a sum greater than 10, knowing these 'best friends' will help us recognise what the units digit of the sum is, and help develop your speed and accuracy.

For example   six plus eight   6 + 8

The answer is greater than 10. Knowing that the best friend of 8 is 2, if we subtract 2 from 6, **the units digit of the answer is 4.**   6 + 8 = **14**

2)   nine plus seven   9 + 7

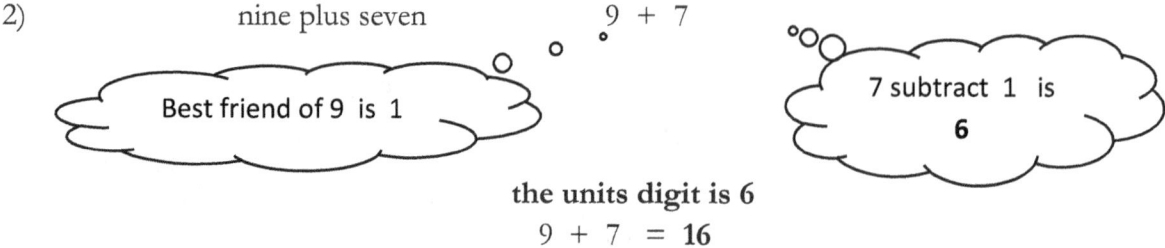

Best friend of 9 is 1

7 subtract 1 is **6**

the units digit is 6
9 + 7 = **16**

Find the missing value of the following sums.

6 + 9 = ___	7 + ___ = 16
8 + 7 = ___	4 + ___ = 11
7 + 5 = ___	___ + 3 = 12
8 + 9 = ___	5 + ___ = 14
5 + 8 = ___	6 + ___ = 13

The second way of improving your speed is to scan the next calculation while you are writing down the answer to the previous calculation. Your brain has a wonderful capacity of doing more than one task at a time.

If you are silently reading this sentence, the 'noise' in your mind is surprisingly loud. Most people people say 'aloud' in their minds each word that they read, this process is called subvocalisation. Thus the speed by which you can solve a calculation, is slowed down by the time it takes you to mentally say a number.

Thus if we mentally perform the following sum,

9 + 8 + 6 + 3 + 8 =

Nine plus eight is seventeen

Seventeen plus six is twenty-three

Twenty-three plus three is twenty-six

Twenty-six plus eight is thirty-four

9 + 8 + 6 + 3 + 8 = 34

While it is difficult to convey a dynamic process on the written page, each step will be outlined.

Remember, the calculation is slowed down by the time it takes you to mentally 'say' a number.

Thus if we mentally add the following numbers

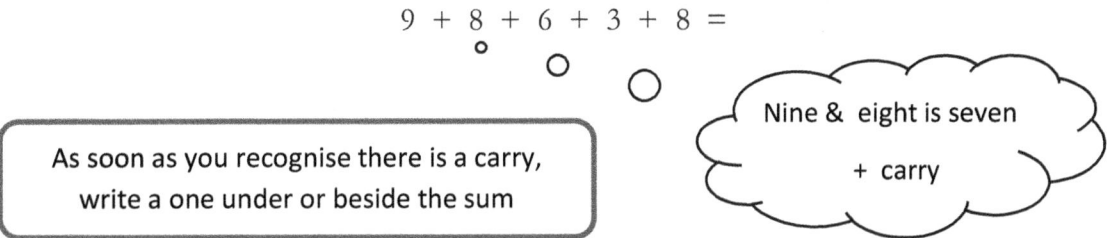

As soon as you recognise there is a carry, write a one under or beside the sum

Nine & eight is seven + carry

We write the carry as a stroke under the pairs of numbers that sum beyond 10, keeping in my mind the number 7.

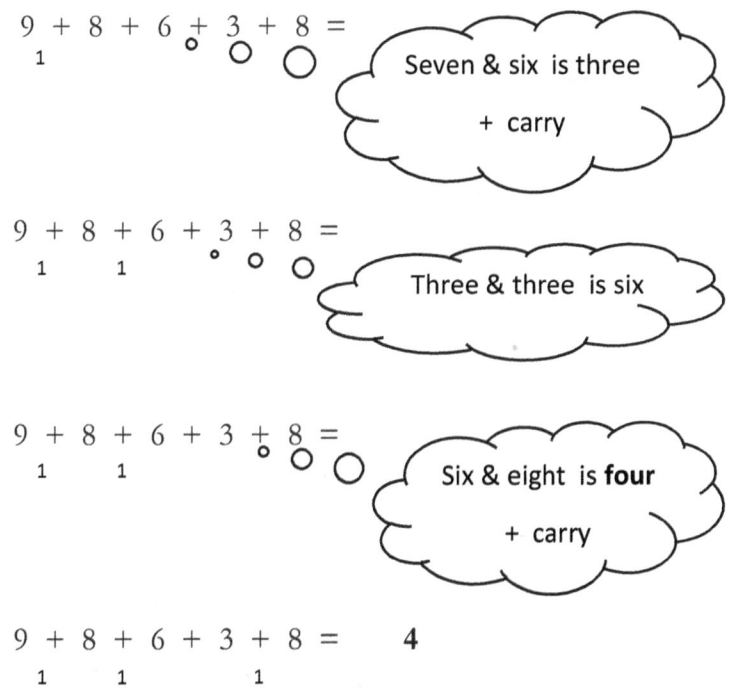

$$9 + 8 + 6 + 3 + 8 = 4$$
$$1 1 1$$

Therefore the units digit of my answer is four, the tens digit is quickly attained by scanning and counting the number of carries.

$$9 + 8 + 6 + 3 + 8 = 3\,4$$

As with anything involving calculations, there is not just one way of doing a sum or calculation.

Identifying the units digit can help quicken the speed of calculations because you do not have to say the entire number to be able to work out a calculation.

Practise questions

1)   8 + 7 + 8 + 9 + 8 =

2)   9 + 3 + 6 + 6 + 9 =

3)   7 + 4 + 7 + 9 + 8 =

4)   7 + 7 + 9 + 3 + 9 =

5)   6 + 3 + 9 + 9 + 7 =

6)   8 +
        7
        9
        5
        8
        3
       ___

7)   6 +
        4
        6
        3
        6
        7
       ___

8)   8 +
        4
        9
        3
        7
        6
       ___

Answers:

1) 40   2) 33   3) 35

4) 35   5) 34   6) 40

7) 32   8) 37

2 digit Addition

The simplicity of only calculating to 10 + the carry, can be extended to 2 digit; 3 digit or beyond additions.

Once again we will use the balloons to represent the processes of the calculations.

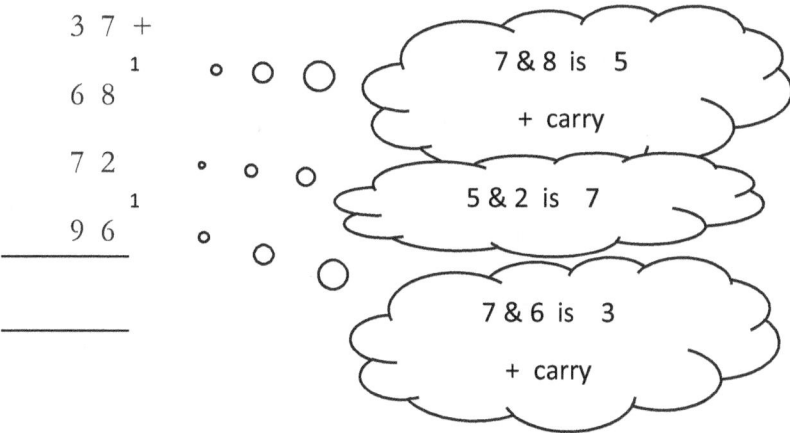

After the units addition the sum should look like this.

```
  3 7 +
     1
  6 8
           The carry 2, is added to the 10s column
  7 2
     1
  9 6
  ___

     3
  ___
```

Performing the tens digit calculations.

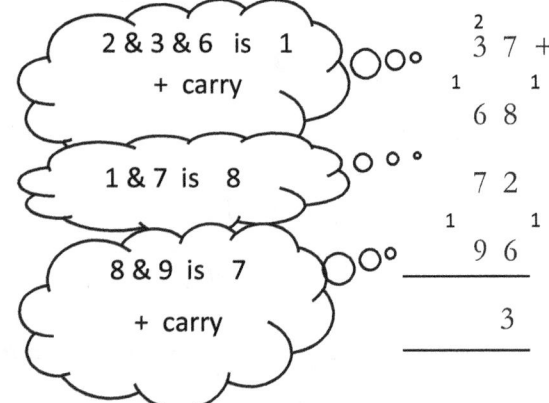

```
         2
       3 7 +
     1     1
       6 8

       7 2
     1     1
       9 6
     _____
           3
```

After the tens addition the sum should look like this.

```
         2
       3 7 +
     1     1
       6 8

       7 2
     1     1
       9 6
     _____
         7 3
```

The carry 2 is added to the 100s column

Therefore the final answer is.

```
         2
       3 7 +
     1     1
       6 8

       7 2
     1     1
       9 6
     _____
       2 7 3
```

If you compare this to the method most of us were taught at school,

The units column (7 + 8 + 2 + 6) adds up to 23,

We say to ourselves, put down the 3 and carry the 2.

Then working in the tens column, (2 + 3 + 6 + 7 + 9) adds up to 27,

We write the 7 in the tens column and the 2 in the hundreds column to calculate the answer 2 7 3.

Working through another example.

```
    7 9 +
         1
    8 4
    9 3
         1
    8 8
    ─────
```

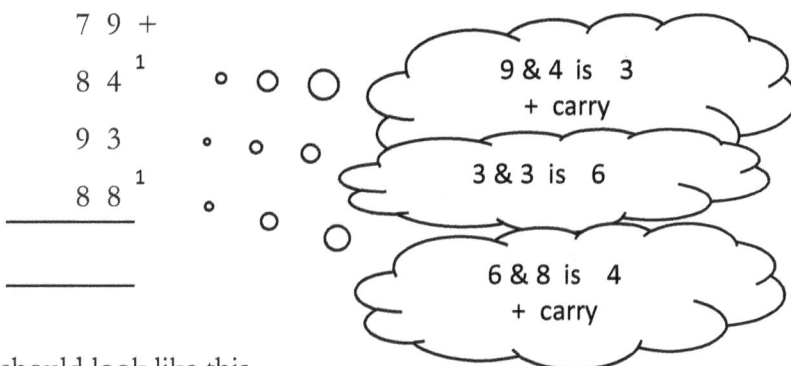

After the units addition the sum should look like this.

```
      2
      7 9 +
           1
      8 4
      9 3
           1
      8 8
      ─────
           4
      ─────
```

The carry 2 is added to the 10s column

Performing the tens digit calculations.

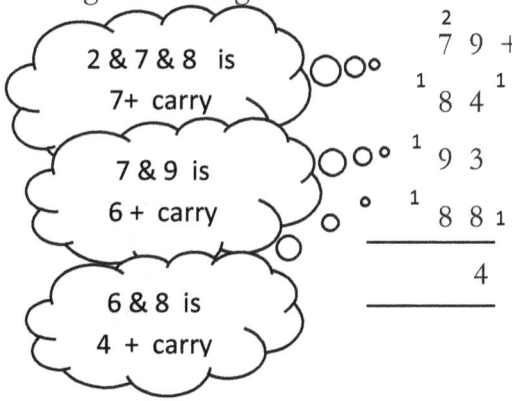

```
      2
      7 9 +
    1      1
      8 4
        1
      9 3
        1
      8 8 1
      ─────
           4
      ─────
```

17

After the tens addition, the sum should look like this.

$$\begin{array}{r} \overset{2}{7}\,9\,+ \\ \overset{1}{8}\,4\,^{1} \\ \overset{1}{9}\,3 \\ \overset{1}{8}\,8\,_{1} \\ \hline 4\,4 \\ \hline \end{array}$$

The carry 3, is added to the 100s column

Therefore the final sum is.

$$\begin{array}{r} \overset{2}{7}\,9\,+ \\ \overset{1}{8}\,4\,^{1} \\ \overset{1}{9}\,3 \\ \overset{1}{8}\,8\,_{1} \\ \hline 3\,4\,4 \\ \hline \end{array}$$

Practise examples

1) 87 +
    83
    99
    47
    ____

2) 98 +
    88
    79
    64
    ____

3) 86 +
    73
    88
    94
    ____

4) 96 +
    33
    99
    84
    45
    ____

5) 86 +
    59
    77
    93
    65
    ____

6) 99 +
    53
    66
    76
    57
    ____

Answers:

1) 316   2) 329   3) 341

4) 357   5) 380   6) 351

Chapter 2

Ahhhh subtraction

"They don't teach subtraction now in the same way they taught us when we were at school!" is often a plaintiff cry for help from parents when faced with the question from their children, "Mum! (or Dad) How do you subtract this?"

Certainly there are differing methods to what happens when the top digit is smaller than its corresponding digit in the bottom row, but if you understand what is happening to the digits and how they are treated then the fear of "trading" vs. "borrow one pay back one" is lessened.

This may be how some parents or adults, who went to school in the 70s or before, were taught how to do subtraction.

$$\begin{array}{r} 9\,8\,3 \\ -\ 5\,6\,4 \\ \hline \end{array}$$ **The 4 units digit is larger than the 3 units digit**

$$\begin{array}{r} 9\,8\,{}^{1}3 \\ -\ 5\,\cancel{6}\,4 \\ 7 \\ \hline \end{array}$$ *If we add ten units to the 3 to make 13, then we need to add 1 to the 6 tens digit to make 7, to balance the equation.*

Effectively we are adding 10 to both lines

Once this has happened we can subtract each digit to get the answer   4 1 9.

In the eighties, the Education authorities decided to change the way subtraction was taught to students, to highlight the direct link between the steps and the mathematical process of subtraction. In Australia, they called this method 'Trading'.

The trading method

The trading method operates to keep the same balance of the number, but only affects the top line.

$9\ ^7\!8\ ^1\!3$ ***We trade 1 tens digit from the 8 to make 7, to add 10 units to the***
$-\ 5\ 6\ 4$ ***3 to make 13.***
              7 tens + 13 units = 83

The answer is still 4 1 9

Trading a ten for 10 units

To demonstrate the trading technique, one of the easiest ways is to use base 10 blocks.

The subtraction   5 4
                  − 3 7
                  ─────

Can be expressed in blocks by the following diagram

When we trade, we trade 1 tens block with tens units block so that each digit is larger than the digit below.

54 - 4 tens   14 units $4\ ^1\!4$
37 - 3 tens   7 units $-\ 3\ 7$
                          ─────
                            1 7

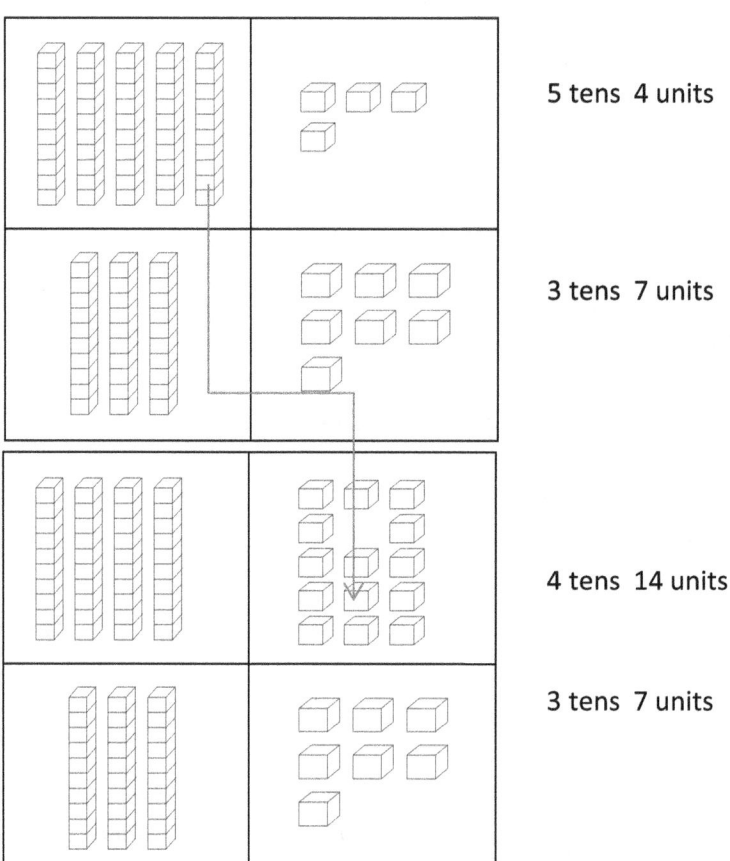

5 tens 4 units

3 tens 7 units

4 tens 14 units

3 tens 7 units

Apply the rule for trading by trading a ten from the following numbers.   **T**ens **U**nits

    i)   6 tens & 3 units (6 3)   becomes   5 tens & 13 units (5 ^{1}3)

    ii)   2 tens & 8 units (6 3)   becomes   1 ten & 18 units   (1 ^{1}8)

    iii) 7 tens & 0 units (7 0)   becomes   6 tens & 10 units (6 ^{1}0)

Try these

    iv)   9 tens & 2 units

    v)   4 tens & eight units

    vi)   1 ten & 3 units

Trading a hundred for 10 tens

$$\begin{array}{r} 4\ 3\ 8 \\ -\ 2\ 6\ 5 \\ \hline \end{array}$$

4 hundreds
3 tens
8 units

2 hundreds
6 tens
5 units

We have to trade 1 hundred for 10 tens

$$\begin{array}{r} 3\ ^13\ 8 \\ -\ 2\ 6\ 5 \\ \hline 1\ 7\ 5 \end{array}$$

3 hundreds
13 tens
8 units

2 hundreds
6 tens
5 units

Apply the rule for trading by trading a hundred from the following numbers.   H T U

    i)   4 hundreds, 5 tens & 2 units   (4 5 2)   becomes   (3 ^{1}5 2)

    ii)   6 hundreds & 8 units   (6 0 8)   becomes   (5 ^{1}0 8)

    iii)   7 hundred   (7 0 0)   becomes   (6 ^{1}0 0)

Try these

    iv)   9 hundreds, 3 tens & 6 units

    v)   5 hundreds, 7 tens & 6 units

    vi)   8 hundreds

Trading both a hundred and a ten.

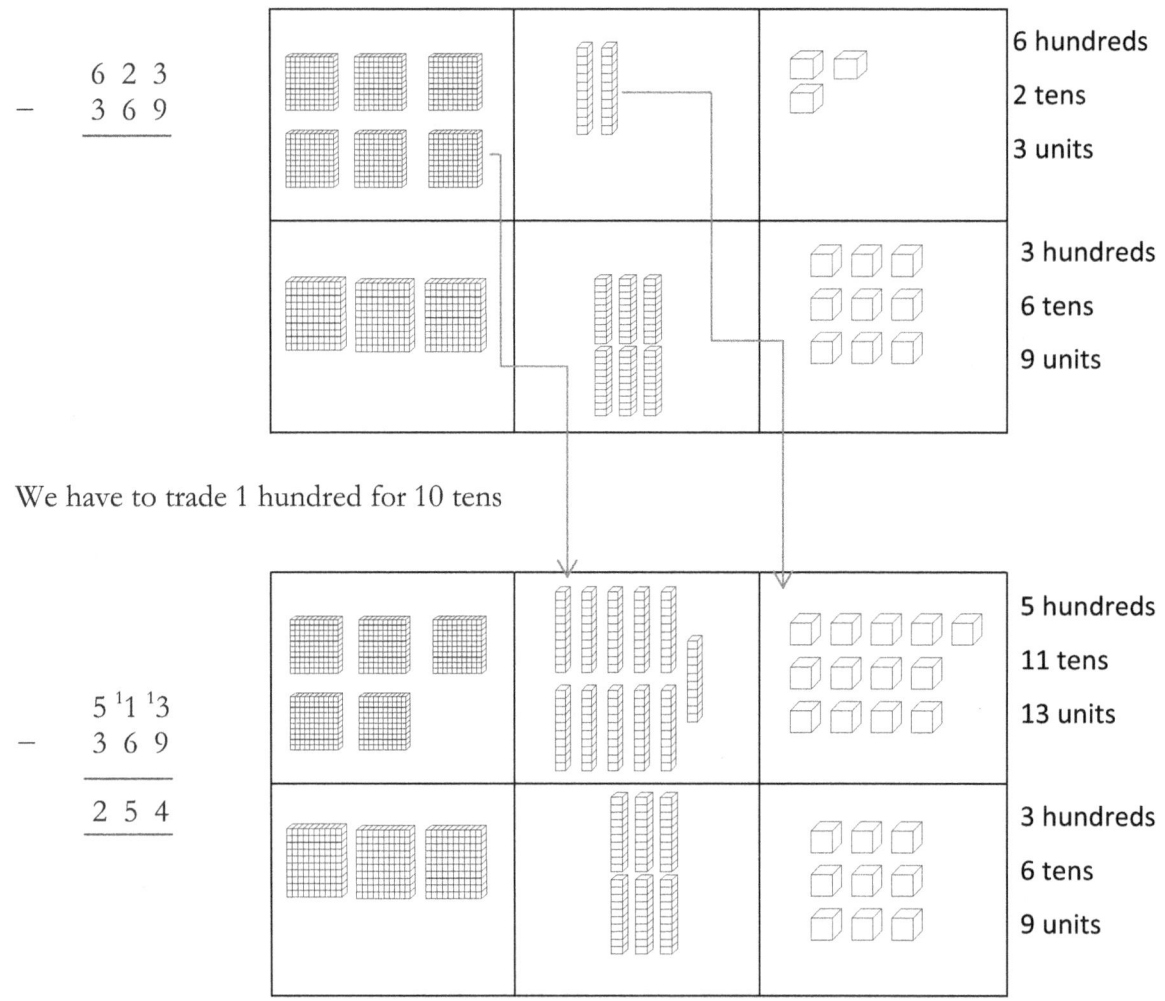

We have to trade 1 hundred for 10 tens

Trade a hundred and a ten in the following.   H   T   U

   i)   7 hundreds, 2 tens & 3 units   (7 2 3)   becomes   (6 ¹1 ¹3)

   ii)   4 hundreds & 9 tens   (4 9 0)   becomes   (3 ¹8 ¹0)

   iii) 9 hundred, 7 tens & 3 units   (9 7 3)   becomes   (8 ¹6 ¹3)

   Try these

   iv)   6 hundreds, 7 tens & 5 units

   v)   5 hundreds, 4 tens & 8 units

   vi)   3 hundreds, 9 tens & 2 units

Subtracting a number from a Base Number: {100s, 1000s 10 000s ...}

One of the best ways to practise the art of trading is subtracting from a number with lots of zeroes. This is relevant to working with money.

e.g.   What is the change from $20, if I purchase an item costing $ 5.45?

Using the same technique that was used to find the difference.

   100 =   9 tens & 10 units

   1 000 =   9 hundreds, 9 tens, & 10 units

   10 000 =   9 thousands, 9 hundreds, 9 tens, & 10 units

   100 000 = 9 ten-thousands, 9 thousands, 9 hundreds, 9 tens, & 10 units

   1 000 000 = 9 hundred-thousands, 9 ten-thousands, 9 thousands, 9 hundreds, 9 tens, & 10 units

We are going to use this pattern to replace the zeroes in our question.

1) 100 becomes $\overset{9\ 1}{\cancel{1}\cancel{0}0}$
     -68 -68
     _____   _____
                                        32

2) $1,000$ becomes $\overset{9\ 9\ 1}{\cancel{1},\cancel{0}\cancel{0}0}$
     -547 -547
     _____   _____
                                        453

3) $\quad\quad\quad\begin{array}{r}10{,}000\\-8{,}243\\\hline\end{array}\quad$ becomes $\quad\begin{array}{r}\overset{9\,9\,9\,1}{\cancel{10{,}000}}\\-8{,}243\\\hline 1{,}757\end{array}$

4) $\quad\quad\quad\begin{array}{r}100{,}000\\-46{,}928\\\hline\end{array}\quad$ becomes $\quad\begin{array}{r}\overset{9\,9\,9\,9\,1}{\cancel{100{,}000}}\\-46{,}928\\\hline 53{,}072\end{array}$

If we have digits other that 1, working from the right the first non-zero digit is reduced by 1

$\quad\quad$ 200 = 1 hundred, 9 tens + 10 units

$\quad\quad$ 500 = 4 hundreds, 9 tens + 10 units

$\quad\quad$ 6,000 = 5 thousands, 9 hundreds, 9 tens, & 10 units

$\quad\quad$ 7,300 = 7 thousands, 2 hundreds, 9 tens, & 10 units

$\quad\quad$ 60,000 = 5 ten-thousands, 9 thousands, 9 hundreds, 9 tens, & 10 units

$\quad\quad$ 74,000 = 7 ten-thousands, 3 thousands, 9 hundreds, 9 tens, & 10 units

1) $\quad\quad\quad\begin{array}{r}400\\-27\\\hline\end{array}\quad$ becomes $\quad\begin{array}{r}\overset{3\,9\,1}{\cancel{400}}\\-27\\\hline 373\end{array}$

2) $\quad\quad\quad\begin{array}{r}5{,}000\\-1{,}829\\\hline\end{array}\quad$ becomes $\quad\begin{array}{r}\overset{4\,9\,9\,1}{\cancel{5{,}000}}\\1{,}829\\\hline -3\,171\end{array}$

3) $\quad\quad\quad\begin{array}{r}82{,}000\\-674\\\hline\end{array}\quad$ becomes $\quad\begin{array}{r}\overset{1\,9\,9\,1}{\cancel{82{,}000}}\\-674\\\hline 81{,}326\end{array}$

Practise exercises

Trade a 10 from the number to make an equivalent value

                                          units
1)   6 2 is equivalent to 5 12   2)   7 8 ⇔

                                          tens
3)   5 5 ⇔   4)   4 8 3 ⇔

Trade a 100 from the number to make an equivalent value

                                        tens
5)   6 1 3 ⇔   5 11 3   6)   5 8 4 ⇔

                                       hundreds
7)   2 3 6 ⇔   8)   4,8 3 2 ⇔

Trade a 100 & a 10 from the number to make an equivalent value

                                  Hundreds   units
9)   2 5 8 ⇔   1 14 18   10)   3 6 2 ⇔

                                       tens
11)   7 1 5 ⇔   12)   7 5 7 4 ⇔

Change the base number to make an equivalent value

                                  Hundreds   units
13)   1,0 0 0 ⇔   9 9 10   14)   4 0 0 ⇔

                                   tens
15)   5,0 0 0 ⇔   16)   1 6 0,0 0 0 ⇔

Find the difference between number and the base number

17)   1 0 0 − 5 9 = 4 1   18)   1 0 0 − 8 6 =

19)   1,0 0 0 − 2 7 1 =   20)   1 0,0 0 0 − 6,8 3 7 =

Answers:

1)	5 12	2)	6 18	3)	4 15	4)	4 7 13
5)	5 11 3	6)	4 18 4	7)	1 13 6	8)	4 7 13 2
9)	1 14 18	10)	2 15 12	11)	6 10 15	12)	7 4 16 14
13)	9 9 10	14)	3 9 10	15)	4 9 9 10	16)	1 5 9 9 9 10
17)	41	18)	14	19)	729	20)	3,163

Examples using the trading method.

1) $74 - 35 =$

We need to trade from the tens digit

$$\begin{array}{r} {}^{6}\!\!\not{7}^{1}4 \\ -3\,5 \\ \hline 3\,9 \end{array}$$

2) $426 - 283 =$

We need to trade from the hundreds digit

$$\begin{array}{r} {}^{3}\!\!\not{4}^{1}2\,6 \\ -2\,8\,3 \\ \hline 1\,4\,3 \end{array}$$

3) $8,576 - 2,730 =$

We need to trade from the thousands digit

$$\begin{array}{r} {}^{7}\!\!\not{8},{}^{1}5\,7\,6 \\ -2,7\,3\,0 \\ \hline 5,8\,4\,6 \end{array}$$

4) $6,370 - 3,492 =$

We need to trade from the thousands, hundreds & tens digit

$$\begin{array}{r} {}^{5}\!\!\not{6},{}^{12}\!\!\not{3}\,{}^{16}\!\!\not{7}{}^{1}0 \\ -3,4\,9\,2 \\ \hline 2,8\,7\,8 \end{array}$$

Another way

There is another way to do subtraction, which requires building up to the upper base number. When we build up, we also add this amount to the number on the top line. One of the advantages of this method is that if there are 2 or more trading steps, we simply bypass the need for trading.

This method connects with the previous section on subtract from numbers with 000s

Let's first look at the technique with single digit subtraction.

1)

```
  1 6
-   9
-----
```

Rather than trading we are going to build up the second term to the upper 0s in this case 10.

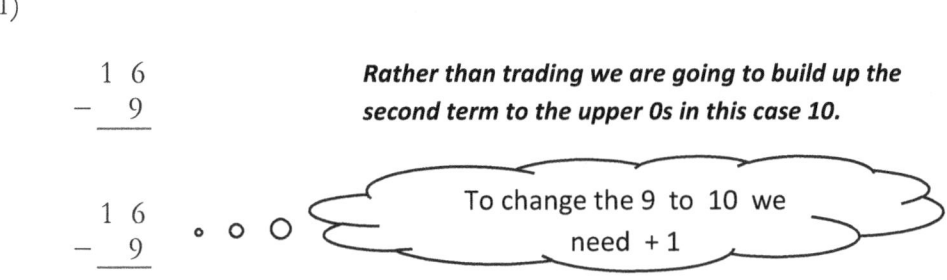

To keep the subtraction balanced, what we do to the bottom line we must do to the top line

$$\begin{array}{r} 1\,6 \;^{+1} \\ -\;\;\;9 \;^{+1} \\ \hline \end{array} \quad \textbf{becomes} \quad \begin{array}{r} 1\,7 \\ -\,1\,0 \\ \hline 7 \end{array} \quad 16 - 9 = 7$$

2)

```
  2 4
-   8
-----
```

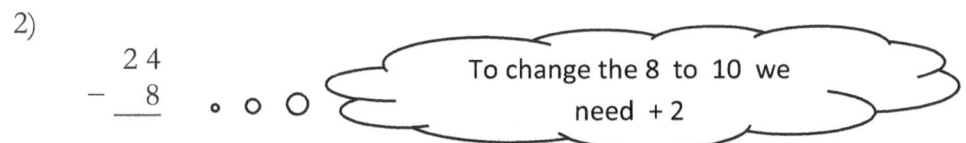

To keep the subtraction balanced, what we do to the bottom line we must do to the top line

$$\begin{array}{r} 2\,4 \;^{+2} \\ -\;\;\;8 \;^{+2} \\ \hline \end{array} \quad \textbf{becomes} \quad \begin{array}{r} 2\,6 \\ -\,1\,0 \\ \hline 1\,6 \end{array} \quad 24 - 8 = 16$$

While there may not be a lot of advantage to do this skill with a single digit, it will be used in harder subtractions and multiplications.

2 digit subtraction

3)

```
  4 2
- 2 7
```

This would require 1 trade so we are going to build up the second term to the upper 0s in this case 30.

```
  4 2
- 2 7
```

To change the 27 to 30 we need + 3

To keep the subtraction balanced, what we do to the bottom line we must do to the top line

```
  4 2  +3        becomes     4 5
- 2 7  +3                  - 3 0
                             1 5
```

$42 - 27 = 15$

4)

```
  1 3 5
-   8 6
```

This requires 2 trades: build up to the upper hundred

To change the 86 to 100 we need + 14

To keep the subtraction balanced, what we do to the bottom line we must do to the top line

```
  1 3 5  +14     becomes     1 4 9
-   8 6  +14               - 1 0 0
                               4 9
```

$135 - 86 = 49$

5)

```
  6 3 2
- 1 7 4
```

This requires 2 trades: build up to the upper hundred

To change the 174 to 200 we need + 26

```
  6 3 2  +26                 6 5 8
- 1 7 4  +26   becomes     - 2 0 0
                             4 5 8
```

$632 - 174 = 458$

3 digit subtraction

6)

```
  6 5 6
−  3 8 7
```

This requires 2 trades: build up to the upper hundred

> To change the 387 to 400 we need + 13

To keep the subtraction balanced, what we do to the bottom line we must do to the top line

```
  6 5 6  +13                        6 6 9
−  3 8 7 +13     becomes          −  4 0 0
                                    2 6 9
```

656 − 387 = 269

7)

```
  1,3 4 2
−    8 6 7
```

This requires 3 trades: build up to the upper thousand

> To change the 867 to 1 000 we need + 133

To keep the subtraction balanced, what we do to the bottom line we must do to the top line

```
  1,3 4 2  +133                     1,4 7 5
−    8 6 7 +133    becomes        − 1,0 0 0
                                     4 7 5
```

1, 342 − 387 = 475

4 & more digit subtraction

8)

```
  4,2 7 6
− 1,6 8 9
```

This requires 3 trading steps: build up to the upper thousand

> To change the 1, 689 to 2, 000 we need + 311

```
  4,2 7 6  +311                     4,5 8 7
− 1,6 8 9  +311    becomes        − 2,0 0 0
                                    2,5 8 7
```

4, 276 − 1, 689 = 2, 587

9) **This requires 4 trading steps: build up to the upper ten thousand**

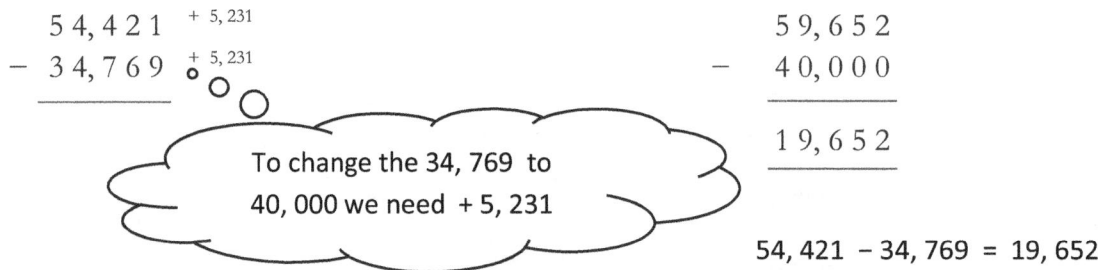

54, 421 − 34, 769 = 19, 652

10) **This requires 4 trading steps: build up to the upper hundred thousand**

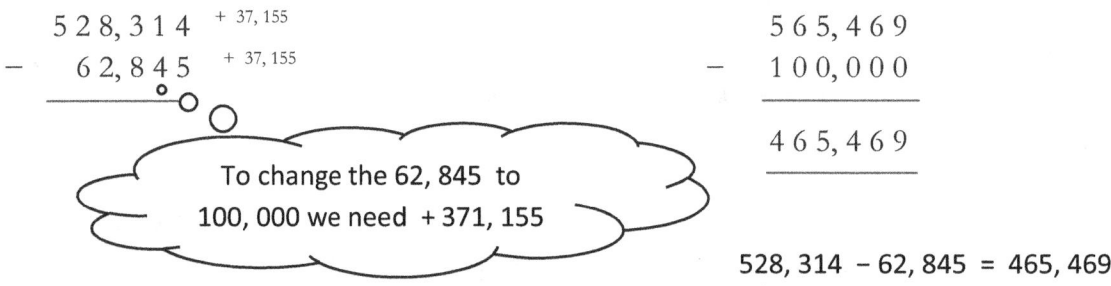

528, 314 − 62, 845 = 465, 469

11) **This requires 3 trading steps: build up to the upper hundred thousand**

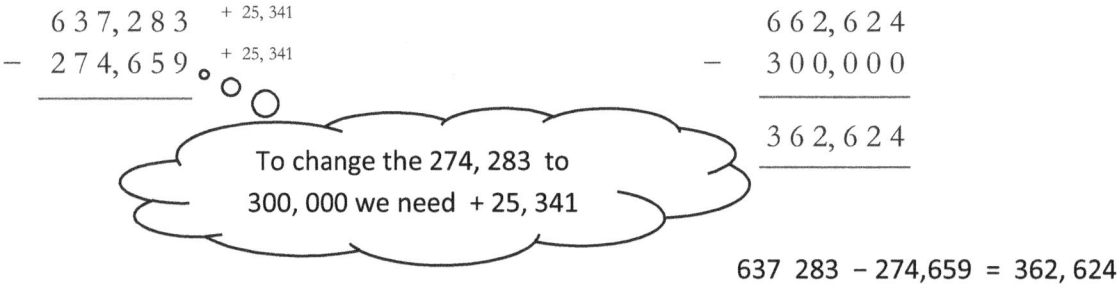

637 283 − 274,659 = 362, 624

This method can be simpler to use, because many find it easier to add than to subtract.

Practise examples on subtraction

1)   6 2
 −   2 3
       ─────

2)   8 1
 −   4 7
       ─────

3)   4 7 2
 −   2 3 9
      ─────

4)   7 6 2
 −   4 2 3
      ─────

5)   5 2 8
 −   1 4 3
      ─────

6)   8 1 6
 −   2 8 3
      ─────

7)   4 2 5
 −   1 7 6
      ─────

8)   9 4 2
 −   7 6 7
      ─────

9)   8, 4 6 2
 −   5, 4 1 8
    ─────

10)   9, 3 6 1
 −   4, 5 2 9
    ─────

11)   2, 3 7 6
 −   1, 4 8 9
    ─────

12)   6, 6 3 4
 −   2, 7 9 8
    ─────

Answers:

1) 39	2) 34	3) 233	4) 339
5) 385	6) 533	7) 249	8) 175
9) 3,044	10) 4,832	11) 887	12) 3,836

Chapter 3

The art of multiplication

How many times have we written out the seven times table, looked at multiplication charts on the bedroom or bathroom doors, maybe played tapes of CD or lyrical songs of times tables, or experienced your teacher reinforcing the numbers again and again and yet ....

For those who know the scoring system in Australian Rules Football (AFL), this story might resonate with you. { 1 goal is 6 points, 1 behind is 1 point }

A teacher in the Year 6 asks the class, "What are seven 7s?" As quick as a flash, Billy says "49". The teacher is a little taken back. Billy has never shown much liking for numbers nor does he answer many Maths questions. "Excellent Billy, now can you tell us what eight 8s are?" Billy replies, "56 Miss."

Billy is the full forward in the Under 13s local AFL team.

Billy may know that 7 goals equal 42 points but he has not made the link that $7 \times 6 = 42$.

In learning the art of multiplication, children move from the concrete (creating of sets or groups: 3 groups of 4 marbles) to the abstract (3×4). Each step along the journey to acquire numerical ability is important and if you miss a step, students may find it difficult to pick up the next one. Some parents complain that students do not know their times tables because they have not learnt them well enough. This is known as the Rote system of learning, where facts are learnt through memorised technique based on repeated sets of writing or saying them out loud.

The approach of multiplication in this book is through patterns rather than by rote learning. It is important to know your multiplications. The method applied in this book is to show you that there exists a pattern not only to find all single digit multiplications, but also to apply these patterns to all multi-digit multiplications.

The two systems of mental mathematics used outlined in this book are the Trachtenberg system for multiplication of single digit numbers and the Vedic Maths system.

The Trachtenberg system and the Vedic Maths system

While some people may have heard of the Vedic Maths system, not many may have heard of Jakow Trachtenberg and the system of mental Mathematics that he devised. Born in Russia in 1888, Jakow Trachtenberg was a mathematician who developed the mental calculation techniques which are now called the Trachtenberg system. After the Russian Revolution of 1917, Trachtenberg fled to Germany. With the rise of the Nazi movement in the 1930, Trachtenberg became critical of their policies. On account of his Jewish faith, he was imprisoned for 7 years in various concentration camps before and during World War II.

In an effort to survive the terrors of the camps, he played with numbers, looking for patterns and devising ways to perform mental calculations quickly. With any scrap of paper he could find, he wrote down the shortcuts for numeracy. After escaping from the camp, he taught this method of quick calculations in the next the few years in Europe before his death. Of course this was the age before the advent of computers. He believed that if children have the ability to add numbers they have the ability to do all kinds of multiplications. The Trachtenberg system uses a system of recognising the tens and units of each product rather than working with the entire number.

In the early part of the 20th century a Hindu scholar and mathematician, Bharati Tirthaji, introduced the Vedic system to the western world. The patterns that are employed date back to an ancient Indian Vedic civilisation that did some sophisticated Mathematics in basic computational numeracy, geometry and algebra. The Vedic system uses patterns of multiplications that can be demonstrated by algebraic methods. It shows that there are many short cuts to multiply numbers like:

88×97

45^2

63×67

17×13

$84 \times 9,999$

and many more.

Terminology

Multiplication is really a short hand form of counting.

When students are learning the art of multiplication, they begin with concrete examples.

If we ask a student to show three groups of four marbles, they might make the following arrangement.

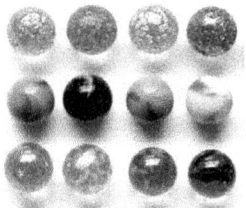

Translating this to a mathematical language   3 lots of 4 marbles

If they were asked them to make four groups of three marbles they might make the following arrangement.

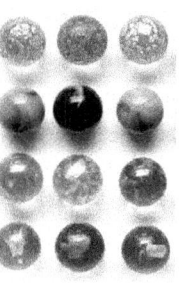

This would mean 4 lots of 3 marbles.

While the total is 12 marbles in each case, how we express the two different examples in mathematical language will make an important distinction

3 lots of 4 marbles   3 × 4   4 lots of 3 marbles   4 × 3

In Mathematics, the result or product is the same, 12 marbles, however, the way we write the multiplication does have a direct relation to something concrete.

To demonstrate this: if three boys have 4 marbles each, or if four boys have 3 marbles each, then while the product 12 is the same, the multiplication order of 3 × 4 or 4 × 3 must have a specific meaning. The **product** in this case is referring to the number of marbles, not the number of boys.

To clarify the multiplication terminology that will be used in this book

$$3 \times 4 = 12$$

Multiplier   **Product**
     Multiplicand

The multiplier is what we are multiplying the sets by (how many times the set is repeated), the multiplicand is the second number (the size of the set) and the product is the result.

Base Numbers. In our decimal system, we use base 10. When I refer to a base number, I am referring to 100, 1 000, 100 000, 1 000 000, etc.

Mental Techniques for multiplication

The easiest technique to learn is multiplication by 11.

When students are asked what does 6 × 11 equal? They easily answer the question - 6 6

In a Year 9 non-calculator numeracy test, Twenty students were given the following question. What does 5 4 × 11 equal? Three students gave the answer

$$54 \times 11 = 5\,544$$

These students learnt at a very early age that when multiplying numbers by 11, all you have to do is write the number twice. What they did not learn was, how multiplication of numbers really works.

Before we look at Trachtenberg's technique, if you YouTube multiplying by 11, you might come across the split and add method.

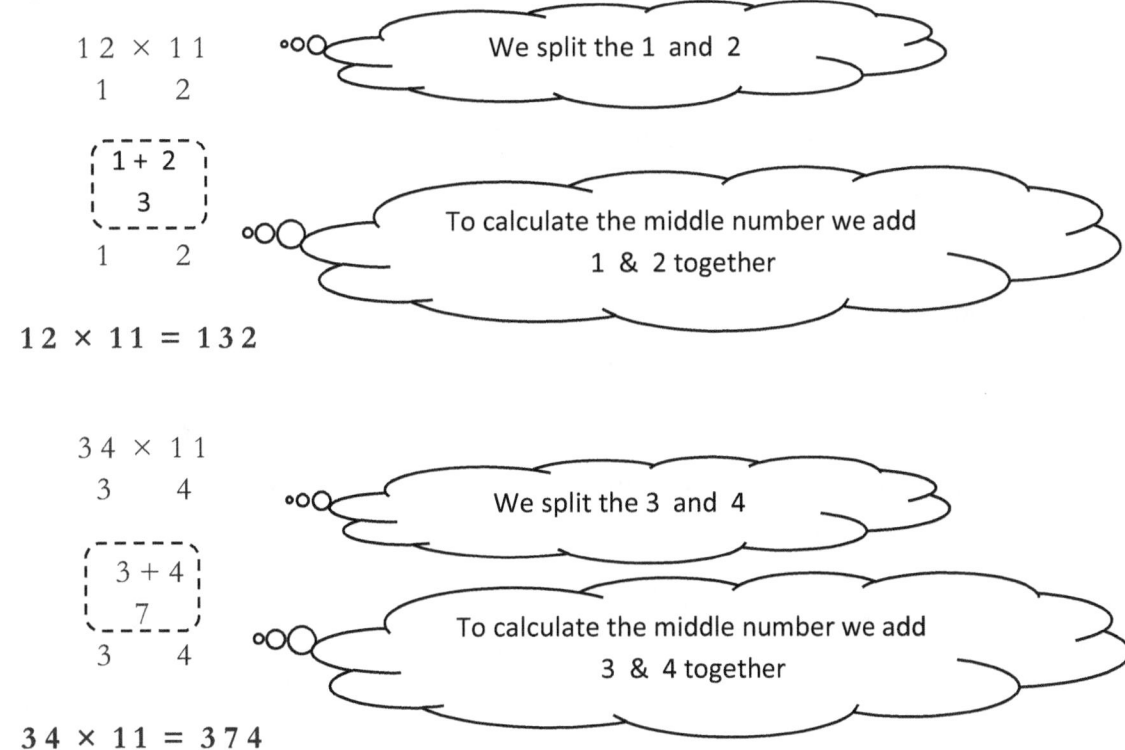

Try the following calculations

54 × 11 =   72 × 11 =   25 × 11 =
594   792   275

If the sum of the two digits is greater than or equal to 10, there will be a carry into the hundreds.

Performing the following calculation the standard way, we can see why this method works.

```
    3 4 ×
    1 1
   ─────
    3 4
  3 4 0
  ─────
  3 7 4
```

If you look at each pair of numbers working down, you can see that the middle number is the sum of the 2 digits 3 & 4

In fact, no matter how big the number, the numbers of the sum of a pair of neighbouring digits. This forms the basis of Trachtenberg's method for multiplying by 11.

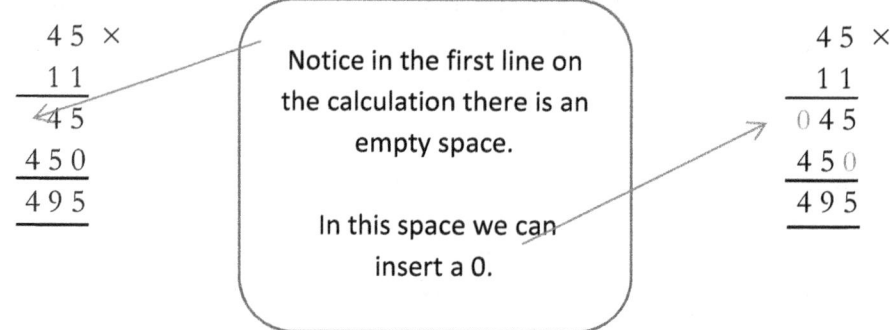

You can see that the two zeroes are at the beginning and the end of the original two digit number.

We are going to use this in most of our multiplication techniques.

When multiplying any number by 11 there are 3 steps. E.g. 6 2 × 1 1

    i.   Write a zero at the beginning and end of the number { ^{0} 6 2 ^{0} }

    ii.   Beginning from the right hand digit, we add its left hand neighbour{ 2 + 0 }

    iii.   If the 2 numbers are equal to or larger than 10, we include a carry in the next sum.

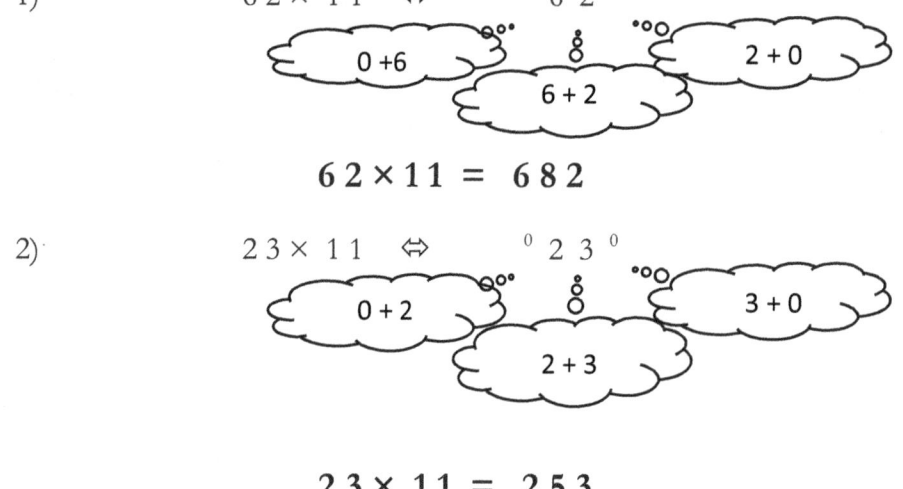

When there is a carry

3)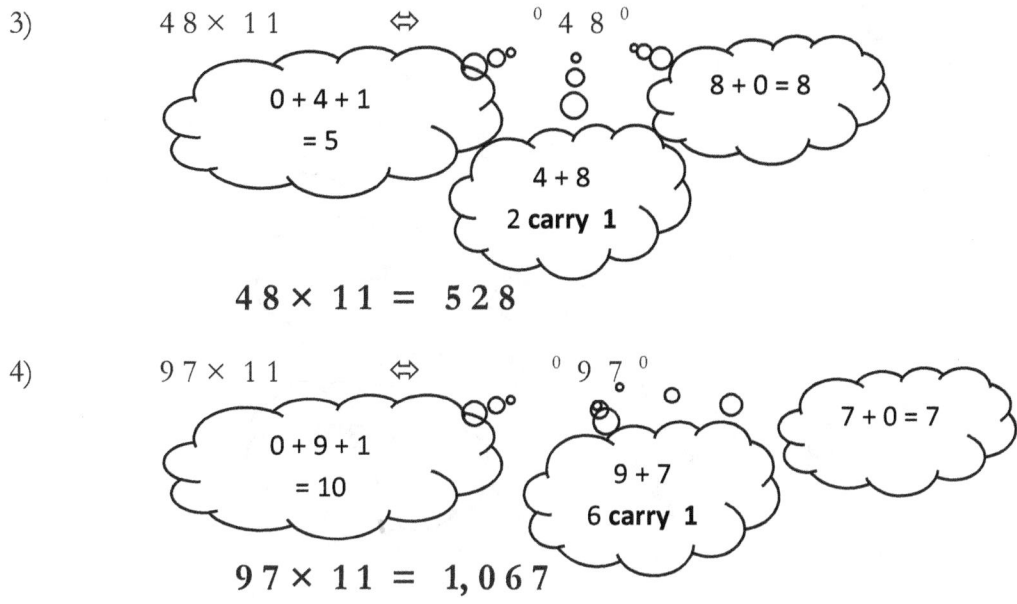

4)

We can extend this to 3 digits multiplied by 11

5)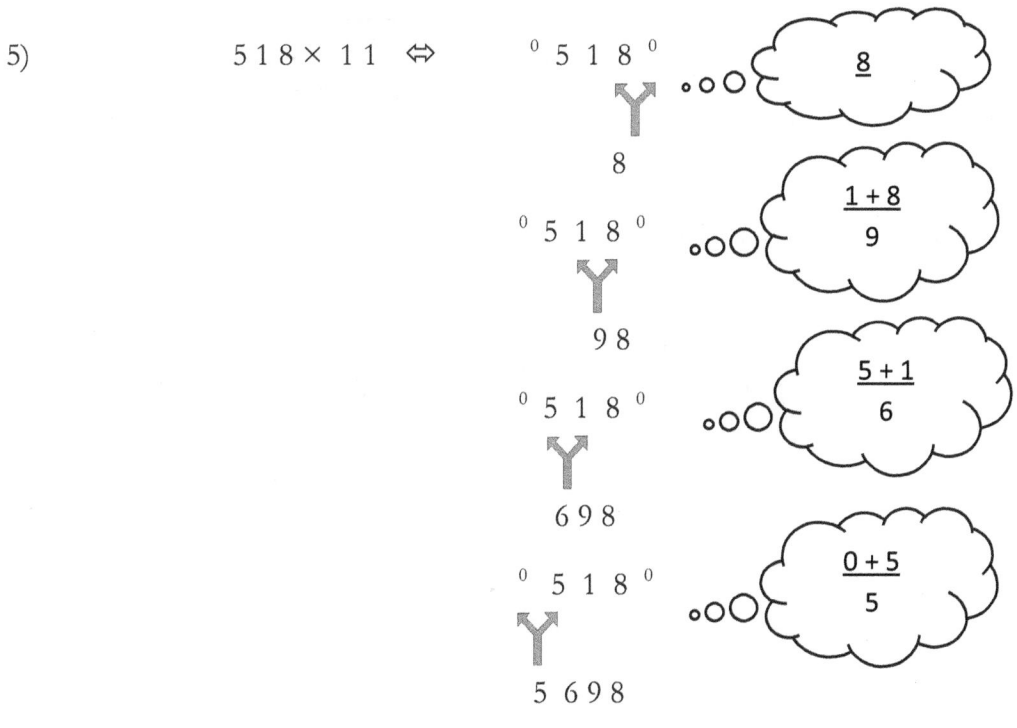

4 digits multiplied by 11

6)   2,4 3 6 × 1 1   ⇔

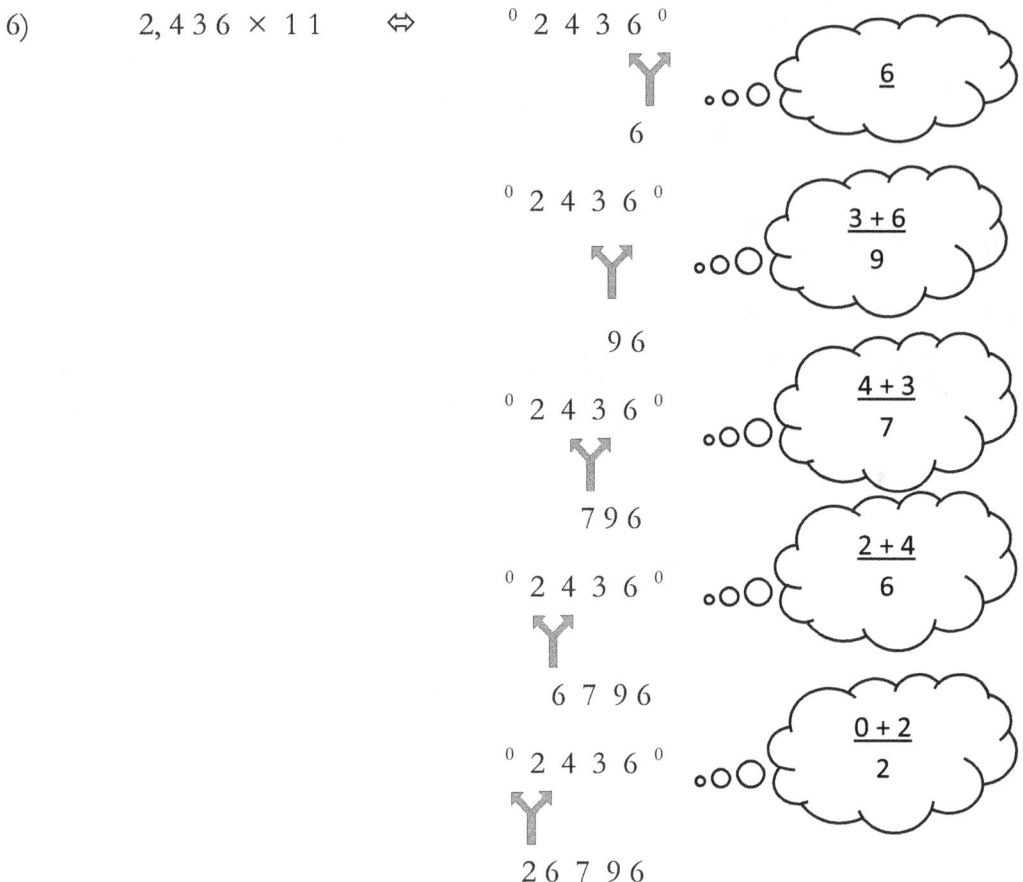

2,4 3 6 × 1 1 = 2 6,7 9 6

3 digits multiplied by 11 with carries

7) $938 \times 11 \Leftrightarrow$ $^0 9\ 3\ 8\ ^0$

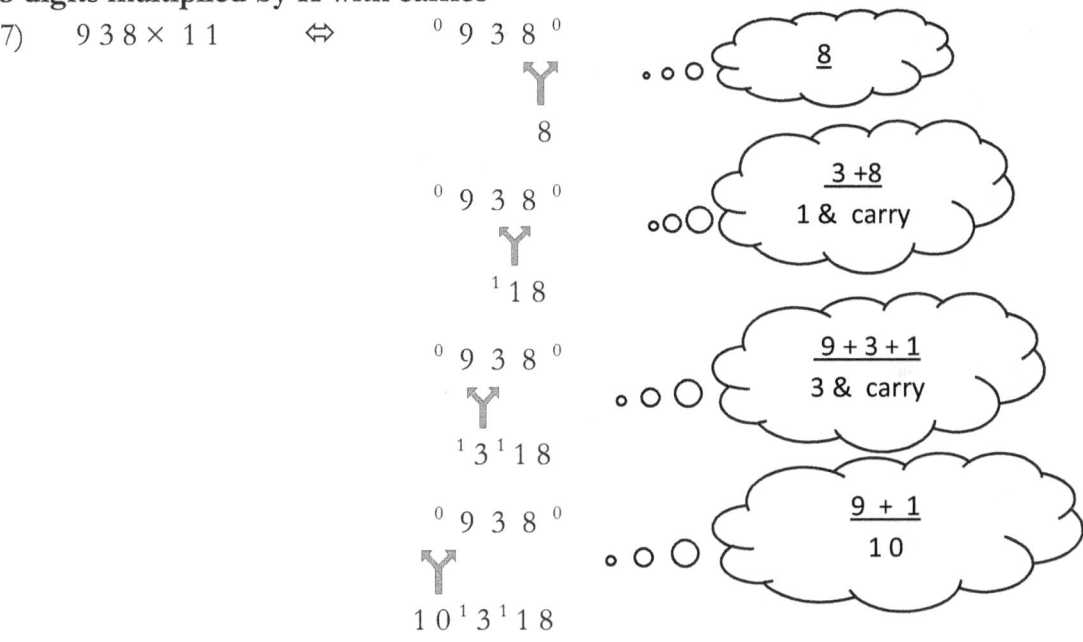

$938 \times 11 = 10,318$

4 digits multiplied by 11 with carries

8) $5749 \times 11 \Leftrightarrow$ $^0 5\ 7\ 4\ 9\ ^0$

$5749 \times 11 = 63,239$

Therefore you can see that no matter how big the questions are, you can amaze yourself and others by multiplying large numbers by 11.

Try that 15 digit number multiplied by 11 from the Introduction Chapter.

$$342,618,032,541,724,351 \times 11$$

$$= 3,768,798,357,958,967,861$$

Multiplication by 11

Practise Questions

1) 15×11 2) 41×11 3) 35×11

4) 38×11 5) 65×11 6) 88×11

7) 91×11 8) 58×11 9) 94×11

10) 307×11 11) 672×11 12) 751×11

13) 233×11 14) 447×11 15) 768×11

16) $2,344 \times 11$ 17) $2,515 \times 11$

18) $9,203 \times 11$ 19) $9,336 \times 11$

Answers:

1)	165	2)	451	3)	385	4)	418
5)	715	6)	968	7)	1,001	8)	638
9)	1,034	10)	3,377	11)	7,392	12)	8,261
13)	2,563	14)	4,917	15)	8,448	16)	25,784
17)	27,665	18)	101,233	19)	102,696		

Multiplying by 9.

If we look at the basic multiplication of 9 times table we get the following

	tens	units
9 × 1	0	9
9 × 2	1	8
9 × 3	2	7
9 × 4	3	6
9 × 5	4	5
9 × 6	5	4
9 × 7	6	3
9 × 8	7	2
9 × 9	8	1

Many students, when asked to write out the 9 times table, construct this beautiful pattern by first writing down the numbers 9 to 1 for the units column, then writing the number 8 up to 0 for the tens column, thus getting each answer correct. But when asked to mentally work out 9 × 7 they may struggle to calculate the answer, or some use their fingers because that was the trick their teacher taught them.

One of the many skills to mastering numbers is looking for patterns. Writing out the table again excluding the tens column we get the following.

9 ×

multiplicand		units
1	⟷	9
2	⟷	8
3	⟷	7
4	⟷	6
5	⟷	5
6	⟷	4
7	⟷	3
8	⟷	2
9	⟷	1

The pattern of the multiplicand and the units is difference between each digit and ten {a digit's best friend}.

The pattern for the tens column.

$9 \times$

multiplicand	tens
1 ↔	0
2 ↔	1
3 ↔	2
4 ↔	3
5 ↔	4
6 ↔	5
7 ↔	6
8 ↔	7
9 ↔	8

The tens digit is always one less than the multiplicand.

Rule for × 9

Thus for a single digit multiplied by 9, we get the following rule:

Units digits: the difference between the multiplicand and ten

Tens digit: subtract one from the multiplicand

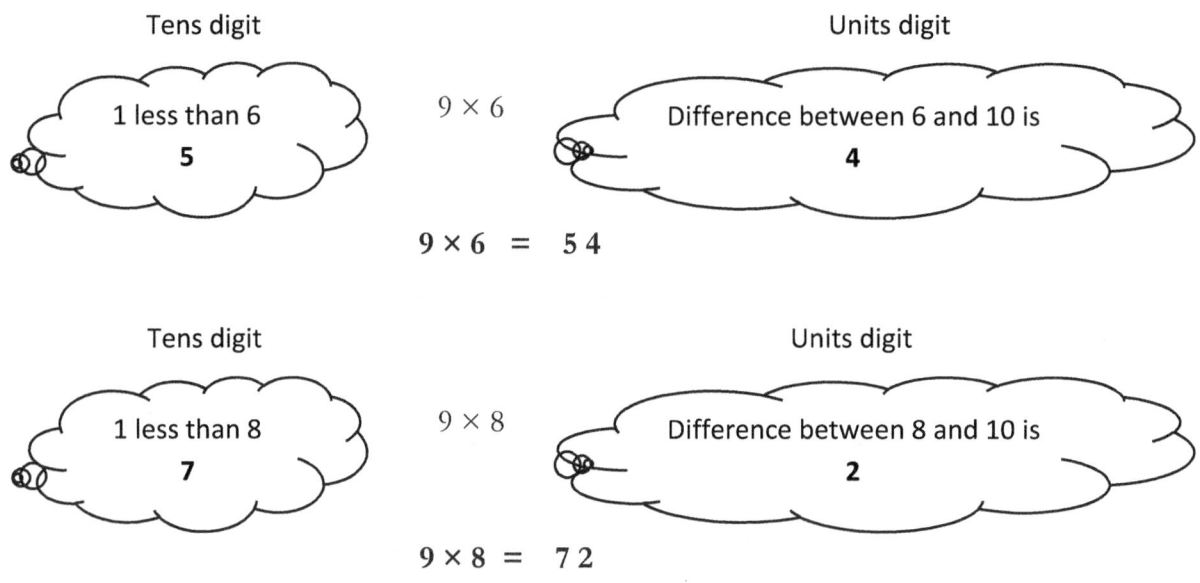

Work though the following tables. (There is no need yet to write the actual answer.)

The units digit
The tens digit   1 less than the number

tens		units		tens		units			answer
7	9 × 8				9 × 7			9 × 3	
	9 × 1				9 × 3			9 × 2	
	9 × 9				9 × 8			9 × 8	
	9 × 6	4			9 × 4			9 × 4	
	9 × 3				9 × 1			9 × 7	
	9 × 5				9 × 6			9 × 5	
	9 × 4				9 × 5			9 × 6	
	9 × 2				9 × 2			9 × 1	
	9 × 7				9 × 9			9 × 9	

There exists a third pattern for multiples of 9. This is extremely useful for checking your calculations (see Chapter 7).

Examining the digits of each multiple of 9, what is another pattern?

tens	units
0 ↔	9
1 ↔	8
2 ↔	7
3 ↔	6
4 ↔	5
5 ↔	4
6 ↔	3
7 ↔	2
8 ↔	1

The sum of the digits adds up to 9. In fact this is true for all multiples of 9.

Ahhhhhhh, but what about $9 \times 11 = 99$?

When we sum the digits of a number we keep summing until we get a single digit.

$99 \longrightarrow 9 + 9 = 18, \longrightarrow 1 + 8 = 9$

General Rule for × 9

When multiplying any number by 9 there are 4 steps. E.g. 84×9

    i.   Write a zero at the beginning and end of the number $^0 8 4 \ ^0$

    ii.   The right hand digit, e.g. '4', we take the difference between it and 10.

    iii.   Then for each digit (working from right to left) we take the difference between it and 9 and add its Right Hand Neighbour RHN (plus a carry).

    iv.   The far left hand digit, we subtract 1 (plus the carry).

Let's see how this works   2 digits multiplied by 9

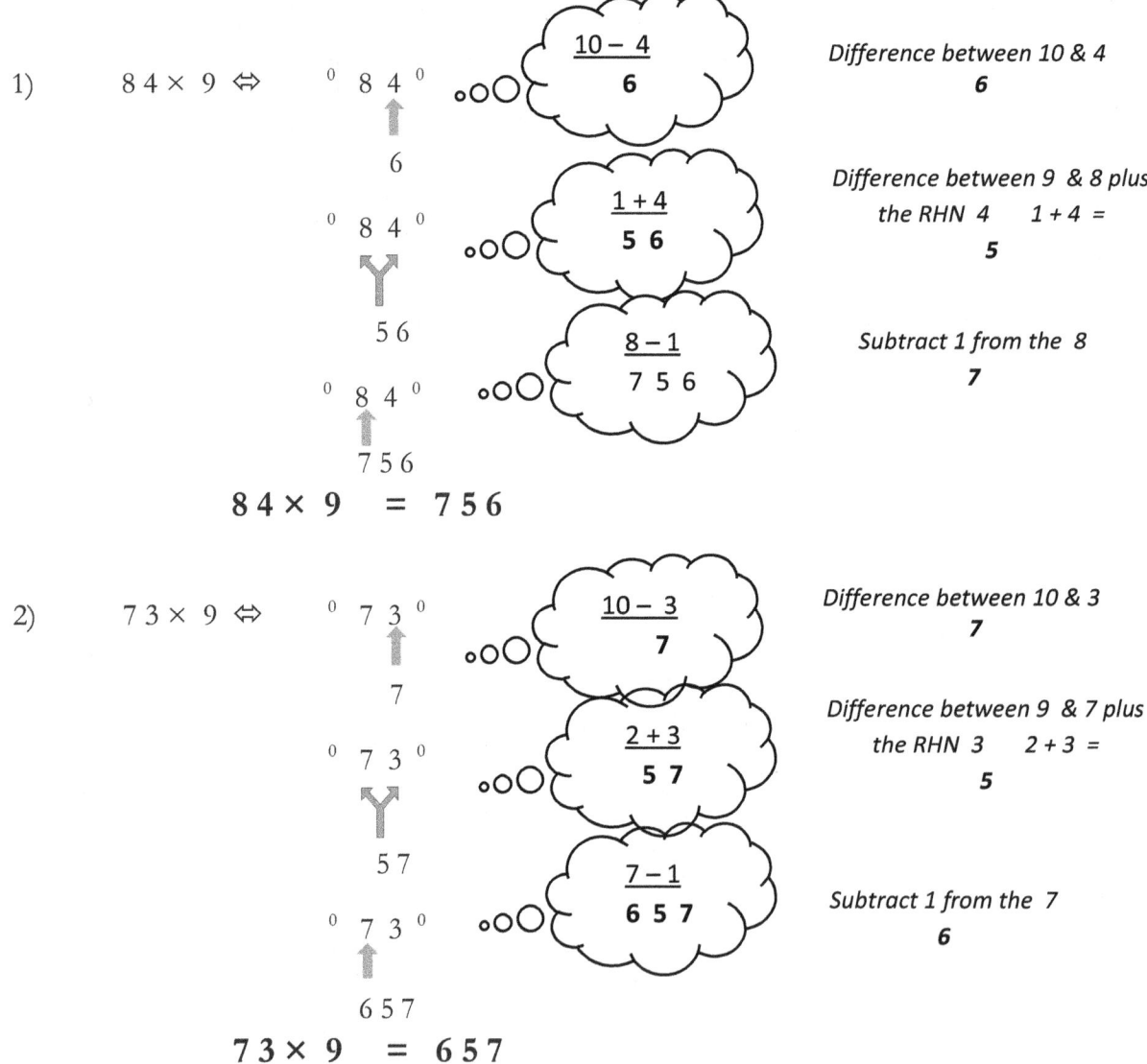

2 digits multiplied by 9 involving a carry

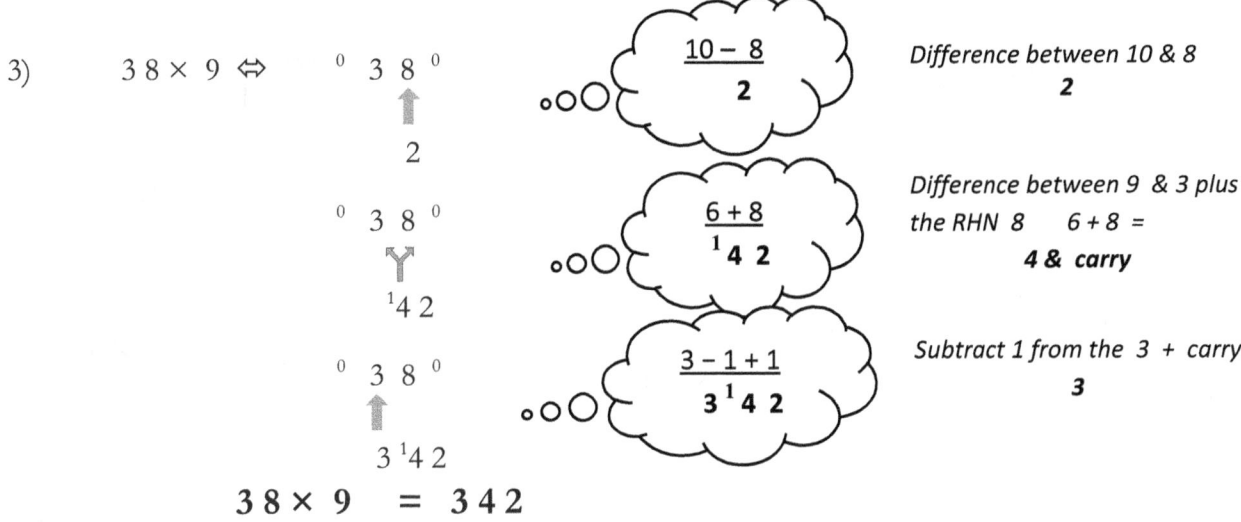

$38 \times 9 = 342$

$49 \times 9 = 441$

As you can see this method does not involve multiplication.

Try the following problems:

1) 72×9 2) 86×9 3) 93×9

4) 36×9 5) 58×9 6) 47×9

Answers:

1) 648   2) 774   3) 837
4) 324   5) 522   6) 423

3 digits multiplied by 9 involving a carry

5)   5 3 6 × 9

        5 3 6 × 9   ⇔   ⁰ 5 3 6⁰
                                    ↑
                                    4

Difference between 10 & 6
4

⁰ 5 3 6⁰
  Y
¹2 4

Difference between 9 & 3 plus the RHN 6   6 + 6 =
2 & carry

⁰ 5 3 6⁰
 Y
8 ¹2 4

Difference between 9 & 5 plus the RHN 3 + carry   4 + 3 + 1 =
8

⁰ 5 3 6⁰
↑
4 8 ¹2 4

Subtract 1 from the 5
4

       5 3 6 × 9 = 4, 8 2 4

6)   7, 2 8 6 × 9 =

7 2 8 6 × 9
    ↑
    4

Difference between 10 & 6
4

7 2 8 6 × 9
   Y
  7 4

Difference between 9 & 8 plus the RHN 6
1 + 6 = **7**

7 2 8 6 × 9
 Y
¹5 7 4

Difference between 9 & 2 plus the RHN 8
7 + 8 = **5** **& carry**

7 2 8 6 × 9
Y
5 ¹5 7 4

Difference between 9 & 7 plus the RHN 2 plus trade   2 + 2 + 1 = **5**

7 2 8 6 × 9
↑
6 5 ¹5 7 4

Subtract 1 from the 7
6

      7, 2 8 6 × 9 = 6 5, 5 7 4

Multiplying by 8.

If we look at the basic multiplication of 8 times table we get the following

	tens	units
8 × 1	0	8
8 × 2	1	6
8 × 3	2	4
8 × 4	3	2
8 × 5	4	0
8 × 6	4	8
8 × 7	5	6
8 × 8	6	4
8 × 9	7	2

There is a pattern with the units digit the is similar to the 9s but with a difference:

8 ×

multiplicand	units
6	8
7	6
8	4
9	2

The units are calculated by the double the difference between the multiplicand and ten.

8 ×

multiplicand	units
6	8
7	6
8	4
9	2

Double Difference between 10 & 6   **8**

$2 \times (10-7) \Leftrightarrow 2 \times 3 =$ **6**

$2 \times (10-8) \Leftrightarrow 2 \times 2 =$ **4**

$2 \times (10-9) \Leftrightarrow 2 \times 1 =$ **2**

With the multiples of 8 from 1 to 5, when we double the difference between the number and 10 we get a carry.

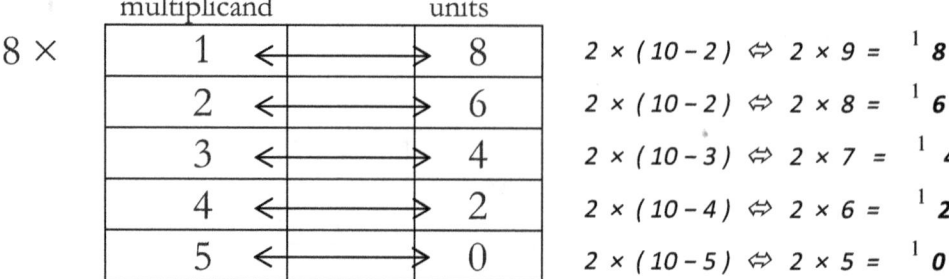

The pattern for the tens column.

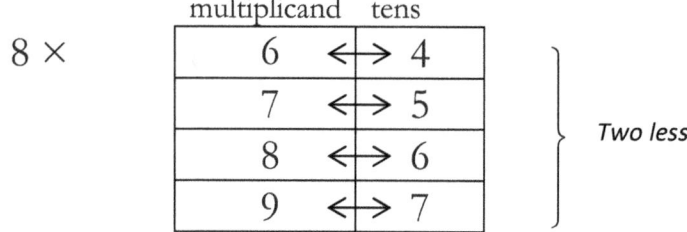

Notice for each number the tens digit is 2 less than the multiplicand.

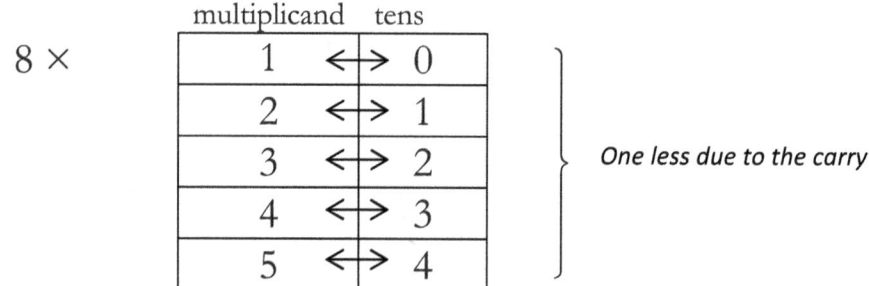

For the multiples for 1 – 5, the carry makes the tens digit 1 less than the number.

Rule for × 8

Thus for a single digit multiplied by 8, we get the following rule:

Units digit: double the difference between the number and ten

Tens digit: two less than the number plus the carry

Let's work through the following examples"

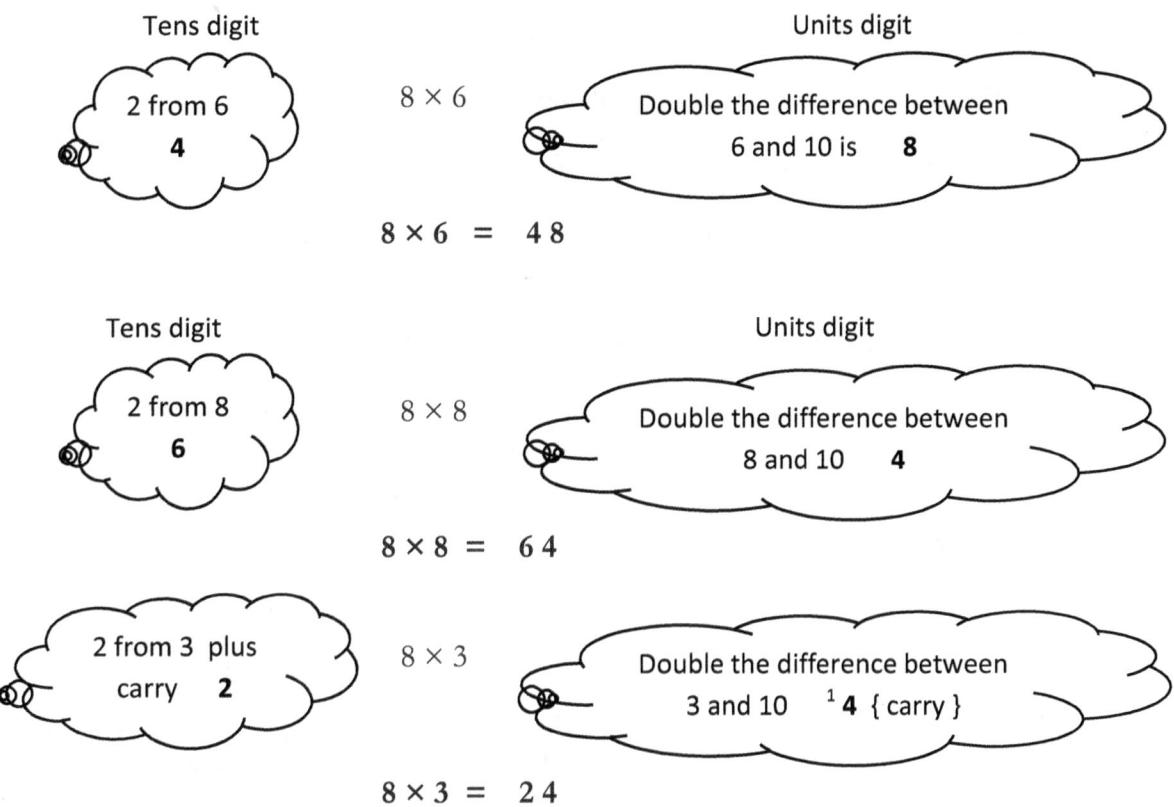

$8 \times 6 = 48$

$8 \times 8 = 64$

$8 \times 3 = 24$

Work though the following tables.

The units digit   Double the he difference between the number and 10

The tens digit   2 less than the number plus the carry

tens		units		tens		units			answer
6	8×8				8×7			8×6	
	8×1				8×3			8×2	
	8×9				8×8			8×8	
	8×6				8×4			8×4	
	8×3	4			8×1			8×3	
	8×5				8×6			8×5	
	8×4				8×5			8×7	
	8×2				8×2			8×1	
	8×7				8×9			8×9	

General Rule for × 8

When multiplying any number by 8 there are 4 steps.   E.g. 67×8

    i.   Write a zero at the beginning and end of the number $^0 6\ 7^{\ 0}$

    ii.   The right hand digit, double the difference between it and 10

    iii.   Then for each digit (working from right to left) we take double the difference between it and 9 and add its Right Hand Neighbour RHN   (plus a carry).

    iv.   The far left hand digit, we subtract 2 from the number plus the carry.

Let's see how this works with 2 digits multiplied by 8.

1) 67×8

$67 \times 8 \quad \Leftrightarrow \quad ^0 6\ 7^{\ 0}$

                             6

Double the difference between 10 & 7 (3)
$2 \times 3 = $ **6**

$67 \times 8 \quad \Leftrightarrow \quad ^0 6\ 7^{\ 0}$

                       $^1 3\ 6$

Difference between 9 & 6 plus the RHN 7
$6 + 7 = $ **3 & carry**

$67 \times 8 \quad \Leftrightarrow \quad ^0 6\ 7^{\ 0}$

                     $5\ ^1 3\ 6$

Subtract 2 from the 6 + carry
$6 - 2 + 1 = $ **5**

$$67 \times 8 \ = \ \mathbf{536}$$

2) 43×8

$43 \times 8 \quad \Leftrightarrow \quad ^0 4\ 3^{\ 0}$

                       $^1 4$

Double the difference between 10 & 3 (7)
$2 \times 7 = $ **4 & carry**

$43 \times 8 \quad \Leftrightarrow \quad ^0 4\ 3^{\ 0}$

                     $^1 4\ ^1 4$

Difference between 9 & 4 plus the RHN 3
plus trade $10 + 3 + 1 = $
4 & carry

$43 \times 8 \quad \Leftrightarrow \quad ^0 4\ 3^{\ 0}$

                 $3\ ^1 4\ ^1 4$

Subtract 2 from the 4 + carry
$4 - 2 + 1 = $ **3**

$$43 \times 8 \ = \ \mathbf{344}$$

As you can see this method also does not involve multiplication.

Try the following problems:

1) 86 × 8   2) 73 × 8   3) 96 × 8

4) 28 × 8   5) 46 × 8   6) 32 × 8

Answers:
1) 688   2) 584   3) 768
4) 224   5) 368   6) 256

The technique for multiplication by 3 or more digits is demonstrated on the web site at www.modmaths.com.au .

Multiplying by 6.

The 6 times table is

	tens	units
6 × 1	0	6
6 × 2	1	2
6 × 3	1	8
6 × 4	2	4
6 × 5	3	0
6 × 6	3	6
6 × 7	4	2
6 × 8	4	8
6 × 9	5	4

Separating the even and odd multiples to explore patterns with the units digit:

6 ×

multiplicand	units
2 ↔	2
4 ↔	4
6 ↔	6
8 ↔	8

When we multiply an even number by 6, the units digit is the same as the number.
Multiplying the odd numbers by 6

6 ×

multiplicand	units
1	6
3	8
5	0
7	2
9	4

$5 + 5 = {}^1 0$ carry
$7 + 5 = {}^1 2$ carry
$9 + 5 = {}^1 4$ carry

The units digit is always 5 more than the number plus the carry, e.g. $9 + 5 = 14 \Leftrightarrow {}^1 4$
The table of the tens digit for 6 times an even number.

6 ×

multiplicand	tens
2	1
4	2
6	3
8	4

The tens digit is half the number.

6 ×

multiplicand	tens
1	0
3	1
5	3
7	4
9	5

} ½ tens digit + carry from the units column

While the relationship is not as easy to see, the same rule applies. For an odd number times 6, when we halve the number, **we ignore the fraction remainder ½** and add the carry to the interger.

6×7 ⇔ units digit $\quad 7 + 5 = {}^1 2$

⇔ tens digit $\quad$ ½ of 7 is $\quad 3 \quad$ {ignore the ½}

plus carry

Rule for × 6

Thus for a single digit multiplied by 6, we get the following rule:

Units digits:

        even   the number,

        odd   the number plus 5;

Tens digit: half the number plus the carry *{ignoring the fraction ½}*.

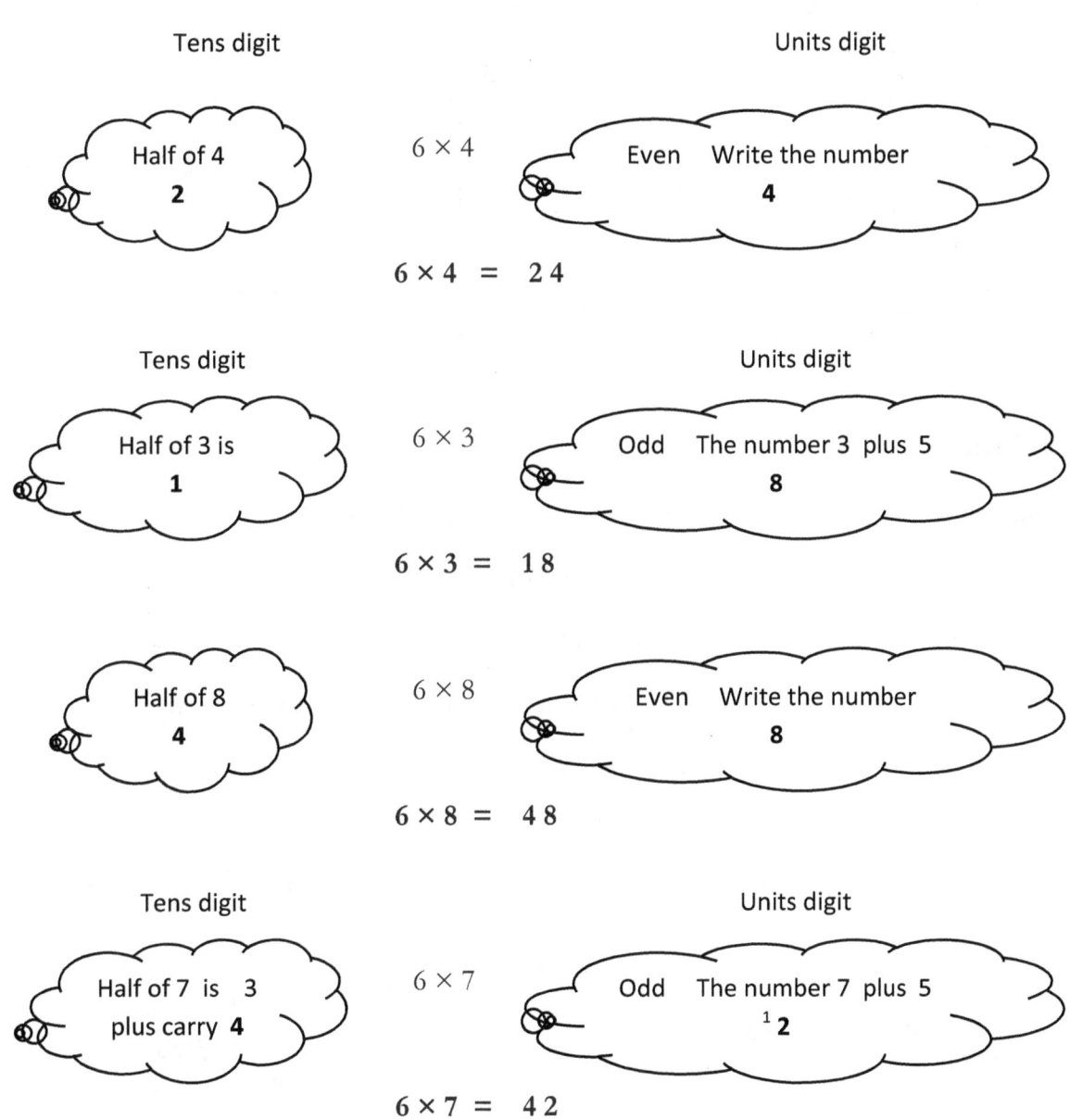

Work though the following tables for multiplying by 6.

 The units digit Even: the number,

  Odd: the number plus 5

 The tens digit halve the number plus the carry

tens	units		tens	units		answer
4	6 × 8			6 × 7		6 × 6	
	6 × 1			6 × 3		6 × 2	
	6 × 9			6 × 8		6 × 8	
	6 × 6			6 × 4		6 × 4	
	6 × 3	8		6 × 1		6 × 3	
	6 × 5			6 × 6		6 × 5	
	6 × 4			6 × 5		6 × 7	
	6 × 2			6 × 2		6 × 1	
	6 × 7			6 × 9		6 × 9	

General rule for × 6

When multiplying any number by 6 there are 2 steps. E.g. 48 × 6

 i. Write a zero at the beginning and end of the number ^{0} 4 8 ^{0}

 ii. If the digit is even: the digit plus half the RHN plus carry **RHN**
  Right Hand Neighbour

 If the digit is odd: the digit plus 5 plus half the RHN plus carry.

Let's see how this works with 2 digits multiplied by 6.

1) 48 × 6

  48 × 6 ⇔ ^{0} 4 8 ^{0} *8 is even*
        ↑   *8*
       8

  48 × 6 ⇔ ^{0} 4 8 ^{0} *4 is even: 4 plus half of 8*
           *8*
       8 8

$$48 \times 6 \quad \Leftrightarrow \quad {}^0\,4\,8\,{}^0$$
$$\uparrow$$
$$2\,8\,8$$

Half of 4
2

$$48 \times 6 \;=\; 288$$

2) 56×6

$$56 \times 6 \quad \Leftrightarrow \quad {}^0\,5\,6\,{}^0$$
$$\uparrow$$
$$6$$

6 is even
6

$$\Leftrightarrow \quad {}^0\,5\,6\,{}^0$$
$$\curlyvee$$
$$3\,6$$

5 is odd: 5 plus half of 6 plus 5
3 + carry

$$\Leftrightarrow \quad {}^0\,5\,6\,{}^0$$
$$\uparrow$$
$$3\,{}^1 3\,6$$

Half of 5 + 1
3

$$56 \times 6 \;=\; 336$$

3) 87×6

$$87 \times 6 \quad \Leftrightarrow \quad {}^0\,8\,7\,{}^0$$
$$\uparrow$$
$${}^1 2$$

7 is odd: 7 plus 5
2 & carry

$$\Leftrightarrow \quad {}^0\,8\,7\,{}^0$$
$$\curlyvee$$
$${}^1 2\,{}^1 2$$

8 is even: 8 plus half of 7 + 1
2 & carry

$$\Leftrightarrow \quad {}^0\,8\,7\,{}^0$$
$$\uparrow$$
$$5\,{}^1 2\,{}^1 2$$

Half of 8 + 1
5

$$87 \times 6 \;=\; 522$$

4)		79 × 6

79 × 6 ⇔ ⁰ 7 9 ⁰		9 is odd: 9 plus 5
                ↑   **4 & carry**
              ¹ 4

        ⇔ ⁰ 7 9 ⁰		7 is odd: 7 plus 5 plus half of 9 + 1
                ⋎   **7 & carry**
              ¹ 7 ¹ 4

        ⇔ ⁰ 7 9 ⁰		Half of 7 + 1
                ↑   **4**
              4 ¹ 7 ¹ 4

79 × 6   =   474

Try the following problems:

1)	46 × 6		2)	76 × 6		3)	69 × 6

4)	23 × 6		5)	54 × 6		6)	37 × 6

	Answers	1)	276	2)	456	3)	414
			4)	138	5)	324	6)	222

Multiplying by 5

The 5 times table is

	tens	units
5 × 1	0	5
5 × 2	1	0
5 × 3	1	5
5 × 4	2	0
5 × 5	2	5
5 × 6	3	0
5 × 7	3	5
5 × 8	4	0
5 × 9	4	5

Separating the even and odd multiplicands:

5 ×

multiplicand	units
2	0
4	0
6	0
8	0

When we multiply an even number by 5, the units digit is always 0
Multiplying the odd numbers by 5

5 ×

multiplicand	units
1	5
3	5
5	5
7	5
9	5

The units digit is always 5
The table of the tens digit for 5 times an even number.

5 ×

multiplicand	tens
2	1
4	2
6	3
8	4

You can see that the tens digit is half the multiplicand.
When we examine the tens digit of the number multiplied by 5.

5 ×

multiplicand	tens
1	0
3	1
5	2
7	3
9	4

Ignore the ½ remainder

For an odd number times, we halve the number & ignore the fraction ½.

Rule for × 5

Thus for a single digit multiplied by 5, we get the following rule:

Units digits: if the digit is even we write 0,   if odd   we write 5

Tens digit: halve the digit ignoring the fraction ½

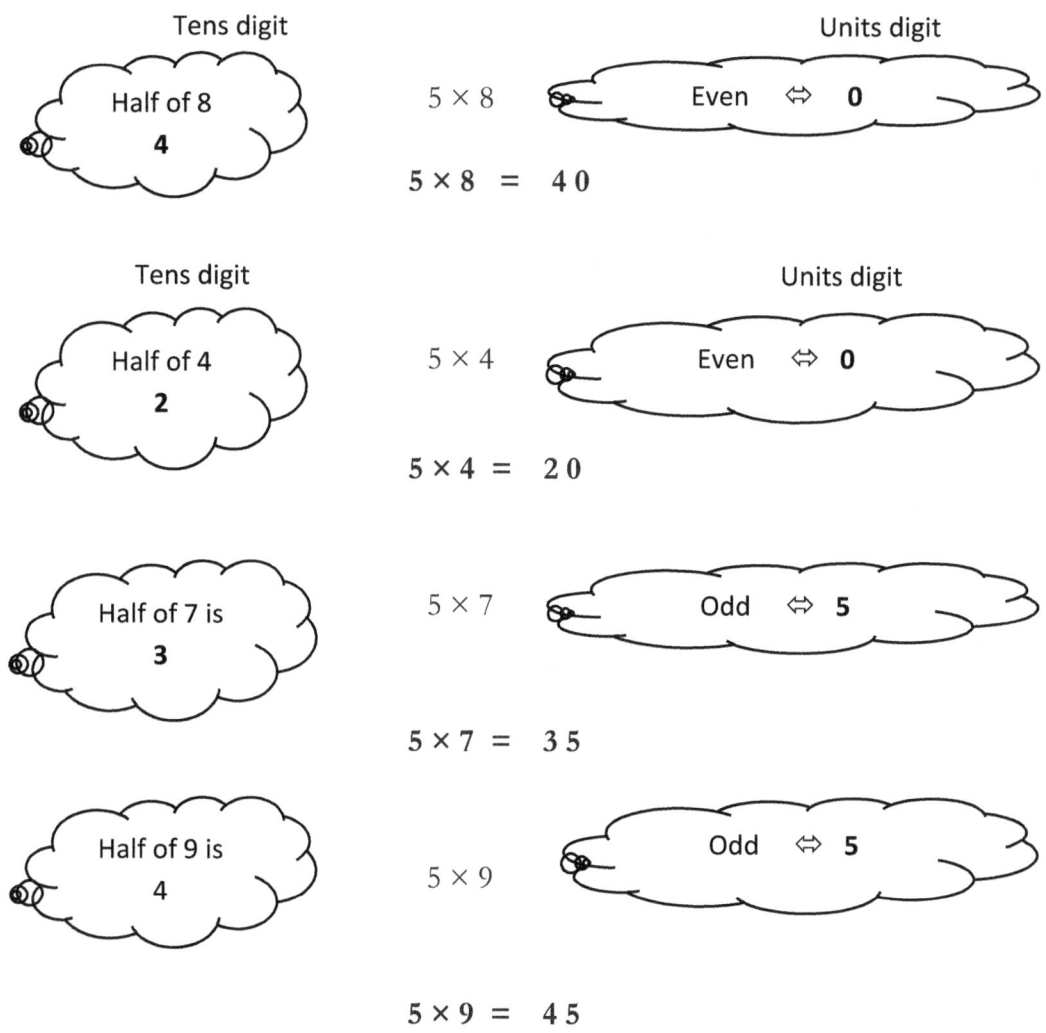

Multiplying by 10 then dividing by 2 gives the same result.

tens	units
4	5 × 8
	5 × 1
	5 × 9
	5 × 6
	5 × 3
	5 × 5	5
	5 × 4
	5 × 2
	5 × 7

tens	units
	5 × 7
	5 × 3
	5 × 8
	5 × 4
	5 × 1
	5 × 6
	5 × 5
	5 × 2
	5 × 9

	answer
5 × 6	
5 × 2	
5 × 8	
5 × 4	
5 × 3	
5 × 5	
5 × 7	
5 × 1	
5 × 9	

General rule for × 5

When multiplying any number by 5 there are 2 steps.   E.g.   6 8 × 5

    i.   Write a zero at the beginning and end of the number ^{0} 6 8 ^{0}

    ii.   If the digit is even: half the RHN

              If the digit is odd: 5 plus half the RHN

Let's see how this works   2 digits multiplied by 5.

1)   6 8 × 5

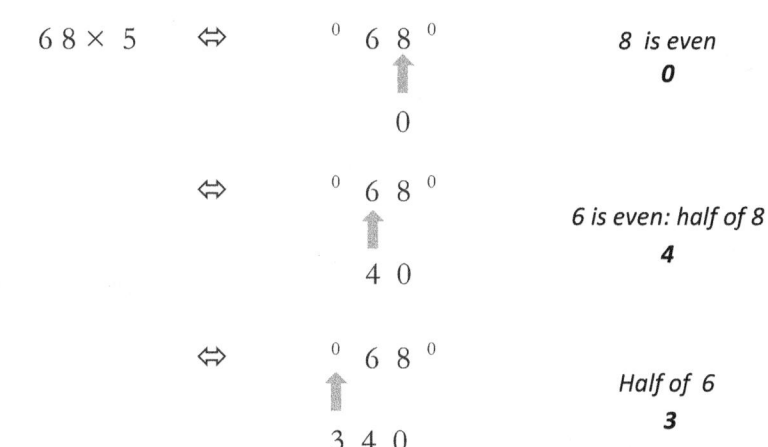

$$6 8 × 5 \;=\; 3 4 0$$

2) 26×5

$26 \times 5 \quad \Leftrightarrow \quad ^0\ 2\ 6\ ^0$

          ↑

         6

6 is even
0

$\Leftrightarrow \quad ^0\ 2\ 6\ ^0$

   ↑

 3 0

2 is even: half of 6
3

$\Leftrightarrow \quad ^0\ 2\ 6\ ^0$

↑

1 3 0

Half of 2
1

$26 \times 5 \ = \ 130$

3) 87×5

$87 \times 5 \quad \Leftrightarrow \quad ^0\ 8\ 7\ ^0$

     ↑

   5

7 is odd:
5

$\Leftrightarrow \quad ^0\ 8\ 7\ ^0$

 ↑

3 5

8 is even: half of 7
3

$\Leftrightarrow \quad ^0\ 8\ 7\ ^0$

↑

4 3 5

Half of 8
4

$87 \times 5 \ = \ 435$

4) 79×5

$79 \times 5 \quad \Leftrightarrow \quad ^0\ 7\ 9\ ^0$

     ↑

   5

9 is odd:
5

$\Leftrightarrow \quad ^0\ 7\ 9\ ^0$

 ↑

9 5

7 is odd: 5 plus half of 9
9

$\Leftrightarrow \quad ^0\ 7\ 9\ ^0$

↑

3 9 5

Half of 7
3

$79 \times 5 \ = \ 395$

Try the following problems:

1)   4 2 × 5   2)   8 4 × 5   3)   4 9 × 5

4)   9 8 × 5   5)   5 7 × 5   6)   2,6 8 4 × 5

Answers:

1) 210   2) 370   3) 245
4) 490   5) 285   6) 13 420

Multiplying by 12

The 12 times table is as follows

	tens	units
12 × 1	1	2
12 × 2	2	4
12 × 3	3	6
12 × 4	4	8
12 × 5	6	0
12 × 6	7	2
12 × 7	8	4
12 × 8	9	6
12 × 9	10	8

Writing out the table again excluding the tens column, we get the following.

12 ×

multiplicand	units
1 ⇐⟶	2
2 ⇐⟶	4
3 ⇐⟶	6
4 ⇐⟶	8
5 ⇐⟶	0
6 ⇐⟶	2
7 ⇐⟶	4
8 ⇐⟶	6
9 ⇐⟶	8

$2 \times 5 = {}^1 0$ ⇔ carry 1
$2 \times 6 = {}^1 2$ ⇔ carry 1
$2 \times 7 = {}^1 4$ ⇔ carry 1
$2 \times 8 = {}^1 6$ ⇔ carry 1
$2 \times 9 = {}^1 8$ ⇔ carry 1

The units digit is always double the multiplicand. When the numbers 5 through 9 are doubled, there is a carry.

The pattern for the tens column.

$12 \times$

multiplicand	tens
1	1
2	2
3	3
4	4
5	6
6	7
7	8
8	9
9	10

+ 1 for carry (from 5 on)

The tens digit is always equal to multiplicand *or from 5 on,* one more than the multiplicand.

Rule for × 12

Thus for a single digit multiplied by 12, we get the following rule:

Units digits: double the number

Tens digit: the number plus the carry

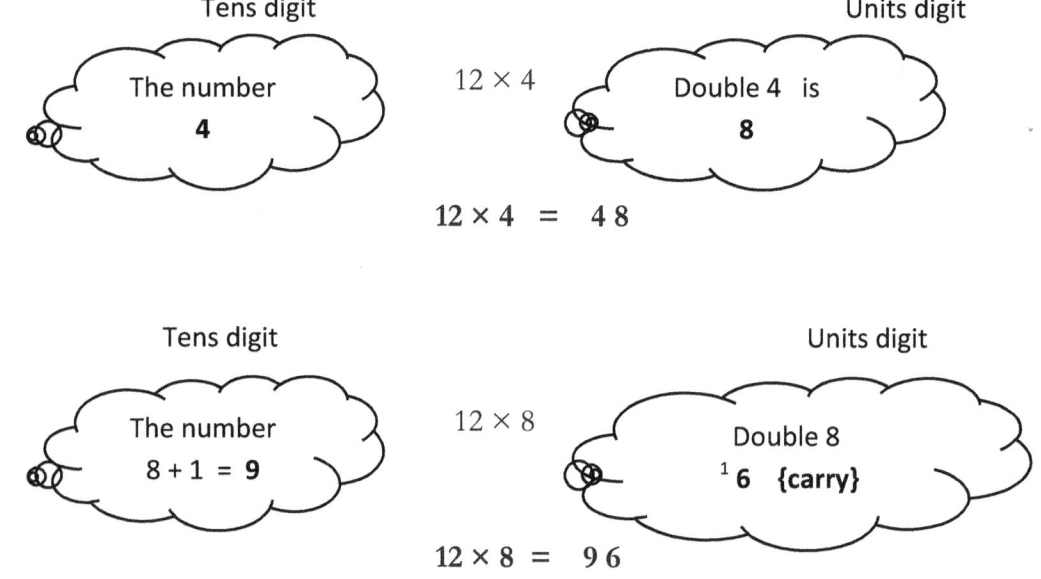

Tens digit: The number 4
12 × 4
Units digit: Double 4 is 8

12 × 4 = 48

Tens digit: The number 8 + 1 = 9
12 × 8
Units digit: Double 8 ¹6 {carry}

12 × 8 = 96

	tens		units
	10	12 × 9	
		12 × 1	
		12 × 8	
		12 × 6	
		12 × 3	
		12 × 5	0
		12 × 4	
		12 × 2	
		12 × 7	

	tens		units
		12 × 7	
		12 × 3	
		12 × 8	
		12 × 4	
		12 × 1	
		12 × 6	
		12 × 5	
		12 × 2	
		12 × 9	

		answer
12 × 6	
12 × 2	
12 × 8	
12 × 4	
12 × 3	
12 × 5	
12 × 7	
12 × 1	
12 × 9	

General rule for × 12

When multiplying any number by 12 there are only 4 steps. E.g. 64×12

    i.   Write a zero at the beginning and end of the number $^0 6\ 4\ ^0$

    ii.   Beginning from the right hand digit, double the digit

    iii.   Double the digit plus the RHN plus the carry

    iv.   The first left hand digit, the number plus the carry

2 digits multiplied by 12

1) 64×12

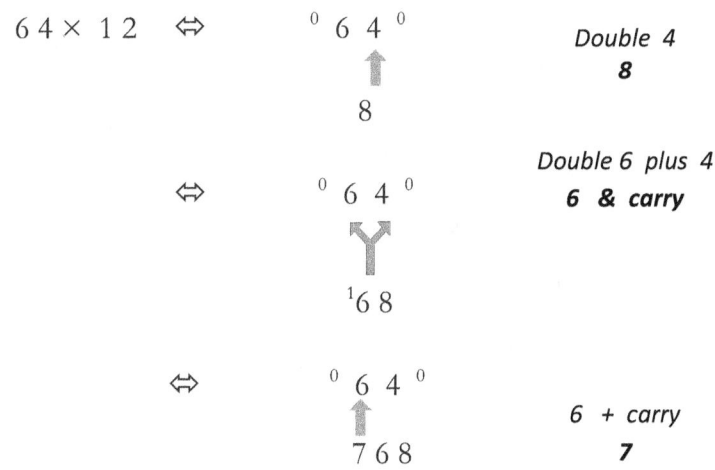

$$64 \times 12 = 768$$

2) 73 × 12

$73 \times 12 \Leftrightarrow$ ⁰ 7 3 ⁰   Double 3
            ↑   **6**
            6

$\Leftrightarrow$ ⁰ 7 3 ⁰   Double 7 plus 3
            Y   14 + 3 = **7 & carry**
           ¹7 6

$\Leftrightarrow$ ⁰ 7 3 ⁰   7 + carry
            ↑   **8**
           8 ¹7 6

73 × 12 = 876

3 digits multiplied by 12

3) 314 × 12

$314 \times 12 \Leftrightarrow$ ⁰ 3 1 4 ⁰   Double 4
                  ↑   **8**
                  8

$\Leftrightarrow$ ⁰ 3 1 4 ⁰   Double 1 plus 4
              Y   2 + 4 = **6**
             6 8

$\Leftrightarrow$ ⁰ 3 1 4 ⁰   Double 3 plus 1
             Y   6 + 1 = **7**
            7 6 8

$\Leftrightarrow$ ⁰ 3 1 4 ⁰
            ↑   0 + 3   **3**
           3 7 6 8

314 × 12 = 3,768

4)   598 × 12

   598 × 12   ⇔ ^{0} 5 9 8 ^{0} ⇑ ^{1}6   Double 8
6 & carry

   ⇔ ^{0} 5 9 8 ^{0} Y ^{1}7^{1}6   Double 9 plus 8 plus carry
18 + 8 + 1 =   **7 & (2) carry**

   ⇔ ^{0} 5 9 8 ^{0} Y ^{2}1^{1}7^{1}6   Double 5 plus 9 plus 2
10 + 9 + 2 =   **1 & (2) carry**

   ⇔ ^{0} 5 9 8 ^{0} ⇑ 7^{2}1^{1}7^{1}6   5 + 2
7

   5 9 8 × 1 2 = 7,1 7 6

4 digits multiplied by 12

5)   6,291 × 12

   6,291 × 12   ⇔ ^{0} 6 2 9 1 ^{0} ⇑ 2   Double 1
2

   ⇔ ^{0} 6 2 9 1 ^{0} Y ^{1}9 2   Double 9 plus 1
18 + 1 =   **9 & carry**

   ⇔ ^{0} $^{2\ 1}$6 2 9 1 ^{0} Y ^{1}4^{1}9 2   Double 2 plus 9 plus 1
4 + 9 + 1 =   **4 & carry**

   ⇔ ^{0} 6 2 9 1 ^{0} Y ^{1}5^{1}4^{1}9 2   Double 6 plus 2 plus 1
12 + 2 + 1 =   **5 & carry**

   ⇔ ^{0} $^{2\ 2\ 1}$6 2 9 1 ^{0} ⇑ 7 ^{1}5^{1}4^{1}9 2   6 + 1
6

   6,2 9 1 × 1 2 = 7 5,4 9 2

Multiplication by 12

Practise Questions

1) 15×12 2) 41×12 3) 35×12

4) 79×12 5) 65×12 6) 88×12

7) 307×12 8) 672×12 9) 751×12

10) $2,344 \times 12$ 11) $4,515 \times 12$

12) $6,203 \times 12$ 13) $8,536 \times 12$

14) $93,263 \times 12$ 15) $6,805,643 \times 12$

Answers:

1) 180   2) 492   3) 420

4) 948   5) 780   6) 1,056

7) 3,684   8) 8,064   9) 9,012

10) 28,128   11) 54,180

12) 74,436   13) 102,432

14) 1,119,156   15) 81,667,716

Multiplying by 7.

The pattern for the 7 times table is a little more complicated. However, as we have already solved all digit multiplications using × 9, × 8, × 6, × 5, the only product left is 7 × 7. The PowerPoints have a more extensive coverage of applying this pattern.

	tens	units
7 × 1	0	7
7 × 2	1	4
7 × 3	2	1
7 × 4	2	8
7 × 5	3	5
7 × 6	4	2
7 × 7	4	9
7 × 8	5	6
7 × 9	6	3

Separating the even and odd multiples to explore patterns with the units digit:

7 ×

multiplicand	units
2 ⟵⟶	4
4 ⟵⟶	8
6 ⟵⟶	2
8 ⟵⟶	6

2 × 6 = $^1 2$ ⇔ carry
2 × 8 = $^1 6$ ⇔ carry

When we multiply an even number by 7, the units digit is double the number.

Multiplying the odd numbers by 7. This pattern is harder to see, and makes using the rule a little more difficult.

7 ×

multiplicand	units
1 ⟵⟶	7
3 ⟵⟶	1
5 ⟵⟶	5
7 ⟵⟶	9
9 ⟵⟶	3

2 × 1 + 5 = 7
2 × 3 + 5 = $^1 1$ ⇔ carry
2 × 5 + 5 = $^1 5$ ⇔ carry
2 × 7 + 5 = $^1 9$ ⇔ carry
2 × 9 + 5 = $^2 3$ ⇔ carry 2

The units digit is double the number plus 5 plus the carry, e.g. 2 × 7 + 5 = 19 $^1 9$

The table of the tens digit for 7 times an even number.

multiplicand	tens
2 ↔	1
4 ↔	2
6 ↔	3
8 ↔	4

7 ×

The tens digit is half the number.

7 ×

multiplicand	tens
1 ↔	0
3 ↔	2
5 ↔	3
7 ↔	4
9 ↔	6

Includes the carry from the units column

For an odd number × 7: halve the number, ***ignore the fraction*** ½ and add the carry.

Rule for × 7.

Thus for a single digit multiplied by 7, we get the following rule:

Units digits:

    even   double the number,

    odd   double the number plus 5;

Tens digit: half the number plus the carry *{ignoring the fraction ½}*.

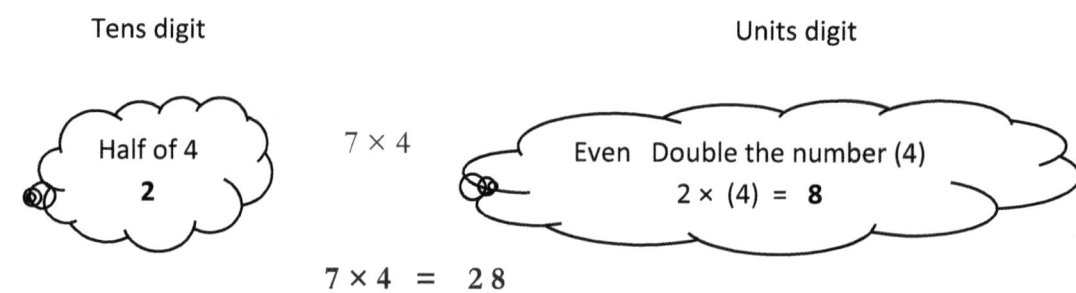

7 × 4 = 28

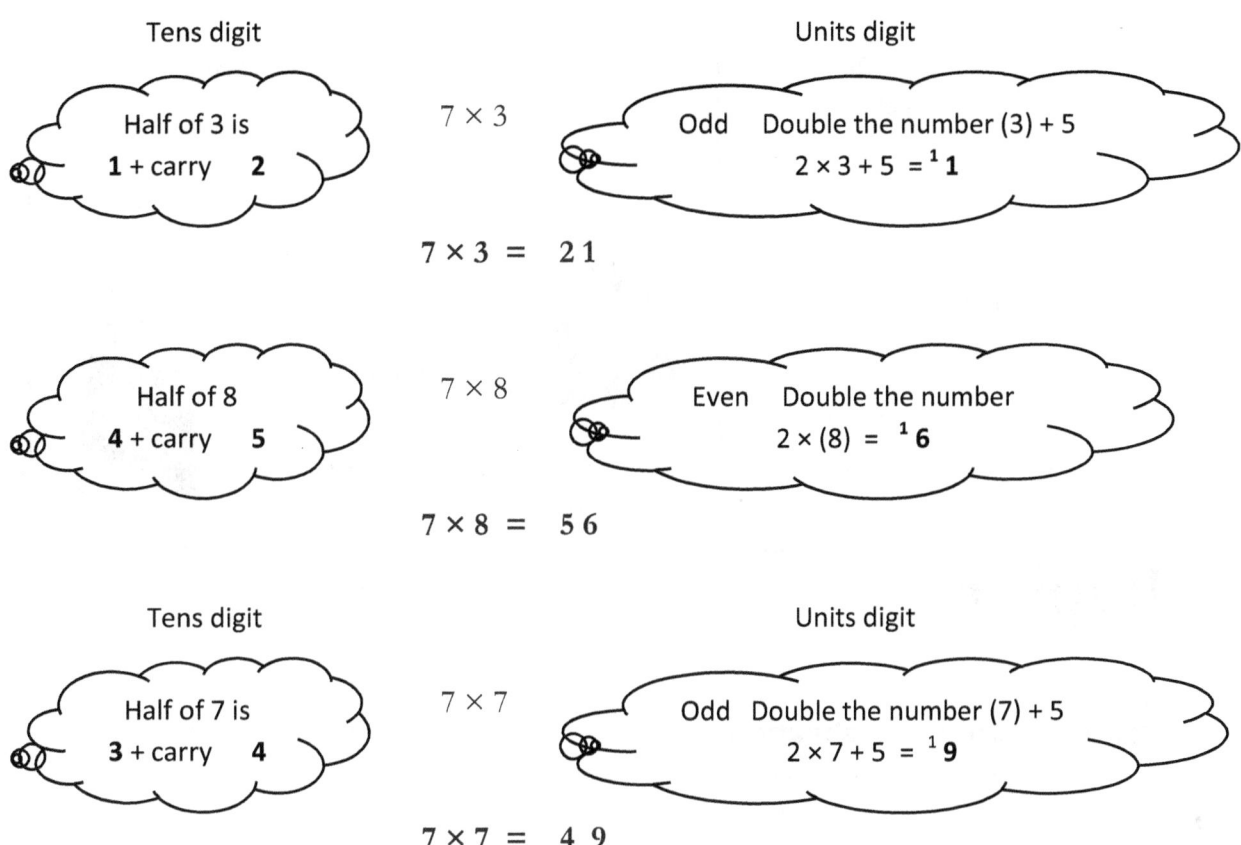

Work though the following tables for multiplying by 7.

The units digit   Even: Double the number   Odd: Double the number plus 5

The tens digit   Halve the number plus the carry

tens		units
5	7 × 8	
	7 × 1	
	7 × 9	
	7 × 7	
	7 × 3	1
	7 × 5	
	7 × 4	
	7 × 2	
	7 × 6	

tens		units
	7 × 6	
	7 × 3	
	7 × 8	
	7 × 4	
	7 × 1	
	7 × 7	
	7 × 5	
	7 × 2	
	7 × 9	

		answer
	7 × 7	
	7 × 2	
	7 × 8	
	7 × 4	
	7 × 3	
	7 × 5	
	7 × 6	
	7 × 1	
	7 × 9	

"What do you mean, it's the wrong kind of right?"

Chapter 4

2 digit by 2 digit multiplication

There are many quick 'tricks' to solve 2 digit by 2 digit multiplications. If you search through *YouTube* you will see the presenter quickly performing a multiplication like 45^2 or 97×96 and many more. While these tricks are handy to know, seldom does the presenter demonstrate a multiplication like 68×57. After I establish the patterns for doing all two-digit multiplications, I will demonstrate these 'tricks' in Chapter 5.

The tried and true, but very cumbersome process, that you may or may not remember (especially when to add that pesky zero) is the following.

$$\begin{array}{r} 97 \times \\ 96 \\ \hline 582 \\ 8{,}730 \\ \hline 9{,}312 \end{array}$$

This process not only involves a number of carries, knowing your times tables, understanding place value to put the digits directly underneath the right place, and especially, understanding why we put that zero in the second working line. After doing these multiplications one then has to add up the two lines to get your final answer.

In our inherited 'English' system of mathematical education, emphasis was placed on a logical approach, which stressed the importance of showing and understanding all the steps involved in a calculation. However, these learnt processes seldom escape from the strictures of, "this is the way I was shown how to do it when I was at school, therefore this is the way I will do it now".

The Vedic Maths system works on a pattern that follows all the steps in the above problem but in a quicker and often simpler way.

Rather than work with numbers, I would like to demonstrate the pattern by replacing the digits with letters.

$$\begin{array}{r} a\ b\ \times \\ c\ d \\ \hline ad\ \ bd \\ ac\ \ bc\ \ 0 \end{array}$$

The product in the 1ˢᵗ column **bd** represents the **units** column. Of course there will often be a carry that will have to be used in the 2ⁿᵈ column (tens).

The sum of the cross multiplication in the 2ⁿᵈ column **ab + bc** represents the **tens** column. Similarly there will often be a carry that will have to be used in the 3ʳᵈ column (hundreds).

The product in the 3ʳᵈ column **ac** represents the **hundreds** column. If the product is 10 or larger, the answer will be in the thousands.

Looking at this pattern, a connecting line is used to represent that the digits are multiplied together.

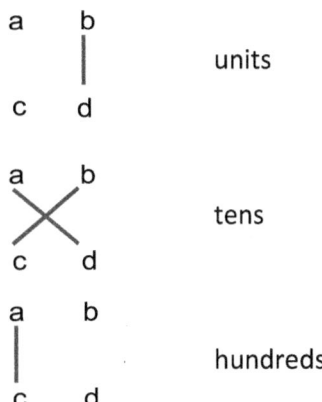

After teaching Mathematics for over 20 years, I know how that 'little' word that starts and ends with an 'a' (involving letters), strikes fear and trepidation into the hearts of many students.

Thus when you replace the letters in the above pattern with dots we get the following pattern.

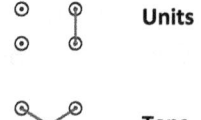 Units

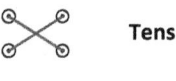 Tens

○ ○
| ⋮
○ ○   Hundreds

Each line connecting the two dots represents a product of two digits.

1)   2 1 × 1 4

```
              2 1 ×
              |
   1 × 4      1 4
      4      ─────
                4
```
○   ○
| ⋮   Units
○   ○

```
              2 1 ×
               ╳
2 × 4 + 1 × 1  1 4
      9      ─────
               9 4
```
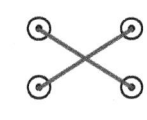 Tens

```
              2 1 ×
              |
    2 × 1     1 4
      2      ─────
              2 9 4
```
○   ○
⋮ | Hundreds
○   ○

2 1 × 1 4 = 2 9 4

2)   2 3 × 2 1

```
              2 3 ×
              |
   3 × 1      2 1
      3      ─────
                3
```
○   ○
| ⋮   Units
○   ○

```
              2 3 ×
               ╳
2 × 1 + 2 × 3  2 1
      8      ─────
               8 3
```
Tens

```
              2 3 ×
              |
    2 × 2     2 1
      4      ─────
              4 8 3
```
○   ○
⋮ | Hundreds
○   ○

2 3 × 2 1 = 4 8 3

72

The next four examples involve a carry.

3) $\quad 36 \times 24$

$$6 \times 4$$
$$\textbf{4 carry 2}$$

$$\begin{array}{r} 3\ 6 \times \\ |\ \\ 2\ 4 \\ \hline ^2 4 \end{array}$$

Units

$$3 \times 4 + 2 \times 6 + 2$$
$$\textbf{6 carry 2}$$

$$\begin{array}{r} 3\ 6 \times \\ \times \\ 2\ 4 \\ \hline ^2 6\ ^2 4 \end{array}$$

Tens

$$3 \times 2 + 2$$
$$\textbf{8}$$

$$\begin{array}{r} 3\ 6 \times \\ |\ \\ 2\ 4 \\ \hline 8\ ^2 6\ ^2 4 \end{array}$$

Hundreds

$$36 \times 24 = 864$$

4) $\quad 42 \times 27$

$$2 \times 7 \quad \textbf{4 carry 1}$$

$$\begin{array}{r} 4\ 2 \times \\ |\ \\ 2\ 7 \\ \hline ^2 4 \end{array}$$

Units

$$4 \times 7 + 2 \times 2 + 1 \quad \textbf{3 carry 3}$$

$$\begin{array}{r} 4\ 2 \times \\ \times \\ 2\ 7 \\ \hline ^3 3\ ^2 4 \end{array}$$

Tens

$$4 \times 2 + 3 \quad \textbf{11}$$

$$\begin{array}{r} 4\ 2 \times \\ |\ \\ 2\ 7 \\ \hline 1\ 1\ ^3 3\ ^2 4 \end{array}$$

Hundreds

$$42 \times 27 = 1{,}134$$

5) $\qquad 59 \times 28$

$\qquad$ $59 \times$
$9 \times 8 \qquad$ **2 carry 7** $\qquad 28$
$\qquad\qquad\qquad\qquad\qquad \overline{^72}$

$\qquad\qquad\qquad\qquad\qquad 59 \times$
$5 \times 8 + 2 \times 9 + 7 \qquad$ **5 carry 6** $\qquad 28$
$\qquad\qquad\qquad\qquad\qquad \overline{^65\,^72}$

$\qquad\qquad\qquad\qquad\qquad 59 \times$
$5 \times 2 + 6 \qquad$ **16** $\qquad 28$
$\qquad\qquad\qquad\qquad\qquad \overline{16\,^65\,^72}$

$$59 \times 28 = 1,652$$

6) $\qquad 63 \times 38$

$\qquad$ $63 \times$
$3 \times 8 \qquad$ **4 carry 2** $\qquad 38$
$\qquad\qquad\qquad\qquad\qquad \overline{^24}$

$\qquad\qquad\qquad\qquad\qquad 63 \times$
$6 \times 8 + 3 \times 3 + 2 \qquad$ **9 carry 5** $\qquad 38$
$\qquad\qquad\qquad\qquad\qquad \overline{^59\,^24}$

$\qquad\qquad\qquad\qquad\qquad 63 \times$
$6 \times 3 + 5 \qquad$ **23** $\qquad 38$
$\qquad\qquad\qquad\qquad\qquad \overline{23\,^59\,^24}$

$$63 \times 38 = 2,394$$

This method will work for all two-digit by two-digit multiplications. There is one drawback, and that happens when the two cross multiplied numbers are large, making the sum of the tens column difficult to mentally compute, as you will see in the next example.

7)   79 × 86

9 × 6	4 carry 5	

```
      7 9 ×
      8 6
     ─────
       ⁵4
```
Units

7 × 6 + 9 × 8 + 5
42 + 72 + 5   119
   9 carry 11

```
      7 9 ×
       ╳
      8 6
     ─────
    ¹¹ 9 ⁵4
```
Tens

7 × 8 + 11   6 7

```
      7 9 ×
      8 6
     ─────
    6 7 ¹¹ 9 ⁵4
```
Hundreds

$$79 \times 86 = 6,794$$

There is another technique that reduces all multiples to either the units or tens digit. A simplified presentation of this technique based on Jakow Trachtenberg's system is given in Chapter 9.

It is important to improve your skills in applying the Vedic Pattern of expansion.

The simplified working can be shown in the following two examples.

i)   1 5 ×
        1 4
       ─────
       1 9 ²0
       ─────
       2 ¹1 0

ii)   2 7 ×
        1 3
       ─────
       2 ¹3 ²1
       ─────
       3 5 1

When you develop your skills, all the working out can be done mentally, thus considerably improving your speed.

Practise questions

1) 13 ×
    12
    ─────

    ─────

2) 16 ×
    14
    ─────

    ─────

3) 19 ×
    17
    ─────

    ─────

4) 14 ×
    18
    ─────

    ─────

5) 23 ×
    16
    ─────

    ─────

6) 26 ×
    18
    ─────

    ─────

7) 23 ×
    29
    ─────

    ─────

8) 38 ×
    27
    ─────

    ─────

9) 76 ×
    39
    ─────

    ─────

10) 78 ×
    57
    ─────

    ─────

Answers:

1) 156   2) 224   3) 323   4) 252
5) 368   6) 468   7) 667   8) 1,026
9) 2,964   10) 4,446

Chapter 5

"How can I trust your information when you're using such outdated technology?"

Shortcuts in Multiplication

While it is essential to understand the pattern to solve any 2 digit by 2 digit multiplications. There are special shortcuts that are easy to learn.

Numbers just above 100

$$104 \times 106$$

The trick to solve this multiplication quickly is finding the difference of each number from 100:

i. Take the difference of each number from 100, and write this difference immediately above the number in a circle.

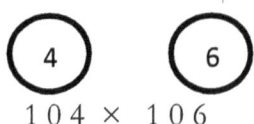

$$104 \times 106$$

ii. The last two digits of the answer are found by multiplying these two circled numbers.

$$4 \times 6 = 24$$

			2	4

iii. The first three digits of my answer are given by cross addition:

$104 + 6 = 106 + 4 = 110$

| 1 | 1 | 0 | 2 | 4 |

$104 \times 106 = 11,024$

2) 105×106

  i.   Take the difference of each number from 100.

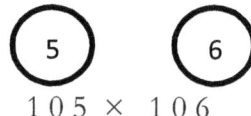

105×106

  ii.   Multiplying these two circled numbers.

$5 \times 6 = 30$

| | | | 3 | 0 |

  iii.   The first three digits of the answer are found by cross addition.

$105 + 6 = 106 + 5 = 111$

| 1 | 1 | 1 | 3 | 0 |

$105 \times 106 = 11,130$

3) 102×104

  i.   Take the difference of each number from 100.

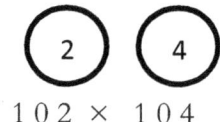

102×104

ii. Multiplying the two numbers in circles.

$$2 \times 4 = 8$$

				8

Remember, as we are working with base 100, we **must** occupy 2 place values.

Thus we put a zero in the tens digit place.

			0	8

iii. The first three digits of my answer is given by cross addition: (both cross additions are the same)

$$102 + 4 = 104 + 2 = 106$$

1	0	6	0	8

$102 \times 104 = 10,608$

If the product of the difference has three digits

e.g. $12 \times 9 = 108$

4) $\qquad 112 \times 109$

i. Take the difference of each number from 100,

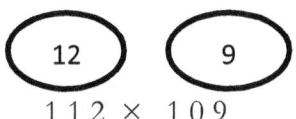

112×109

ii. Multiplying these two circled numbers..

$$12 \times 9 = {}^{1}08$$

If the product has 3 digits then we write the units and tens digits in their correct place and carry over the hundreds digit.

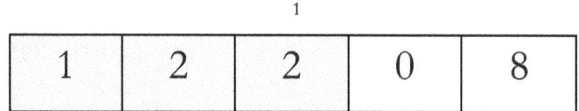

iii. The first three digits of the answer are found by cross addition:

$$112 + 9 + \text{carry} = 121 + 1$$
$$= 122$$

112 × 109 = 12,208

Practise questions

1)   103 × 102   2)   106 × 103

3)   111 × 112   4)   110 × 117

5)   103 × 111   6)   106 × 112

7)   111 × 118   8)   116 × 115

Answers:

1) 10,506   2) 10,918   3) 12,432   4) 12,870

5) 11,433   6) 11,872   7) 13,098   8) 13,340

Numbers just below 100

To demonstrate the method when two numbers are close to 100 but not smaller than 88, let's examine the example at the start of Chapter 4.

$$97 \times 96$$

We know the answer to this multiplication is $\qquad$ 9,312

The trick to solve this multiplication quickly is similar to our first shortcut, except the difference is negative

   i.   Take the difference of each number from 100, and write this difference immediately above the number in a circle.

$$\underset{97 \times 96}{\boxed{-3} \quad \boxed{-4}}$$

   ii.   The last two digits of the answer are found by multiplying the two numbers in circles {multiplying two negatives together makes a positive}.

$$-3 \times -4 = 12$$

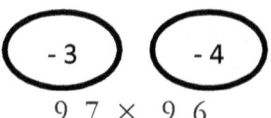

   iii.   The first two digits of my answer is given by cross subtraction: (both cross subtractions are the same)

$$97 - 4 = 96 - 3 = 93$$

| 9 | 3 | 1 | 2 |

$97 \times 96 = 9,312$

2) $\qquad\qquad 95 \times 94$

   i.   Take the difference of each number from 100, and write this difference immediately above the number in a circle.

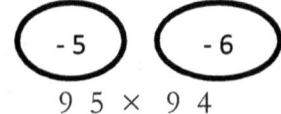

ii. Multiplying these two circled numbers {multiplying two negatives makes a positive}.

$$-5 \times -6 = 30$$

		3	0

iii. The first two digits of the answer are found by cross subtraction.

$$95 - 6 = 94 - 5 = 89$$

8	9	3	0

95 × 94 = 8,930

The algebraic explanation why this works can be downloaded from *modmaths.com.au*

There are two types of multiples where there has to be a slight tweak in the answer. On account that we are working in base 100, the final 2 digits are given by the multiplying the numbers in the circles. What happens if the product of two numbers only has 1 digit? e.g. 2 × 3 or 3 digits e.g. 12 × 9

3)   98 × 97

     i.   Take the difference of each number from 100.

$$(-2) \quad (-3)$$
$$98 \times 97$$

     ii.   Multiplying these two circled numbers.

$$-2 \times -3 = 6$$

			6

But, as we are working with base 100, we must occupy 2 place values

Thus we must put a zero in the tens digit place.

		0	6

iii. The first two digits of my answer is given by cross subtraction: (both cross subtractions are the same)

9 8 − 3   = 9 7 − 2 = 9 5

9	5	0	6

9 8 × 9 7 = 9, 5 0 6

If the product of the differences has three digits?

e.g. - 1 1 × - 1 2 = ¹3 2

4)   8 9 × 8 8

i. Take the difference of each number from 100, and write this difference immediately above the number in a circle.

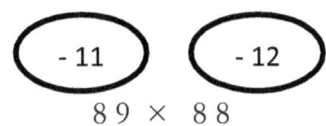

8 9 × 8 8

ii. Multiplying these two circle numbers {multiplying two negatives makes a positive}.

- 1 1 × - 1 2   =   1 3 2

If the product has 3 digits then we write the units and tens digits in their correct place and carry over the hundreds digit.

¹

		3	2

iii. The first two digits of the answer are found by cross subtraction:

89 − 12 + carry = 77 + 1 = 78

| 7 | 8 | 3 | 2 |

89 × 88 = 7,832

Practise questions

1) 93 × 92   2) 96 × 94

3) 99 × 97   4) 94 × 88

5) 93 × 96   6) 96 × 89

7) 91 × 85   8) 81 × 84

Answers: 1) 8,556   2) 9,024   3) 9,603   4) 8,272
           5) 8,928   6) 8,544   7) 7,735   8) 6,804

Numbers on either side of 100

1) 104×97

When finding the differences from 100, one will be positive and the other negative. The product of a positive and negative number is always negative. This will mean we need to require a trade.

    i.   Write this difference immediately above the number in a circle.

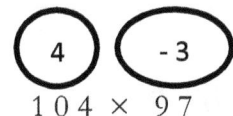

$$104 \times 97$$

    ii.   The last two digits of the answer are found by multiplying the two numbers in circles.

$$4 \times -3 = -12$$

We trade 100   -1   100 − 12

                                                                    88

			8	8

    iii.   The first two (or three) digits of the answer are found by:

$104 - 3 = \quad 97 + 4 \quad = \quad 101$ minus the trade (1)

$$101 - 1 \quad = \quad 100$$

-1

1	0	0	8	8

$$104 \times 97 = 10,088$$

2) 105×92

    i.   Write this difference immediately above the number in a circle.

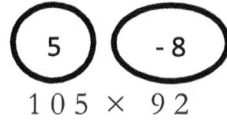

$$105 \times 92$$

    ii.   Multiplying these two circled numbers.

$$5 \times -8 = -40$$

We trade 100   -1   100 − 40

                                                                    60

		6	0

iii. The first two (or three) digits of the answer are found by:

$105 - 8 = \quad 92 + 5 \quad = \quad 97$ minus the trade (1)

$ 97 - 1 \quad = \quad 96$

_{- 1}

9	6	6	0

$105 \times 92 = 9,660$

3) $\qquad\qquad 112 \times 91$

When the product of the two differences is greater than 100, then the trade is from the upper hundred. In the following example we trade 2 (200) from the cross addition.

i. Write this difference immediately above the number in a circle.

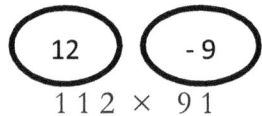
112×91

ii. Multiplying these two circled numbers.

$\qquad\qquad 12 \times -9 \quad = \qquad -108 \qquad\qquad 200 - 108$
$\qquad\qquad\qquad\qquad\qquad\qquad\qquad\qquad\qquad\qquad\qquad 92$

We trade 200

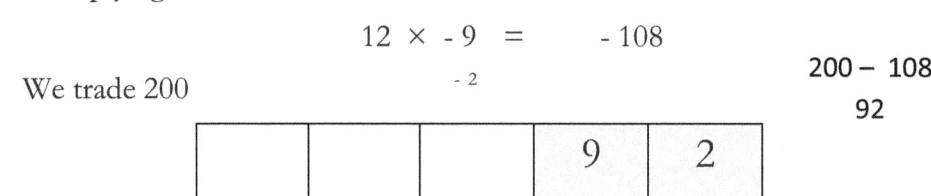

iii. The first two (or three) digits of the answer are found by:

$112 - 9 = \quad 91 + 12 \quad = \quad 103$ minus the trade (2)

$ 103 - 2 \quad = \quad 101$

_{- 2}

1	0	1	9	2

$112 \times 91 = 10,192$

Practise questions

1) 98×102 2) 96×103

3) 110×92 4) 91×103

5) 103×85 6) 96×108

7) 112×91 8) 88×115

Answers:

1) 9,996 2) 9,888 3) 10,120 4) 9,373

5) 8,755 6) 10,368 7) 10,192 8) 10,120

Multiplying the teen numbers.

When multiplying two numbers that are between 10 and 20 (I will call these the teen numbers) there is a simple technique.

1) $\qquad 13 \times 11 = 143$

Separating the product **143** into its tens and units

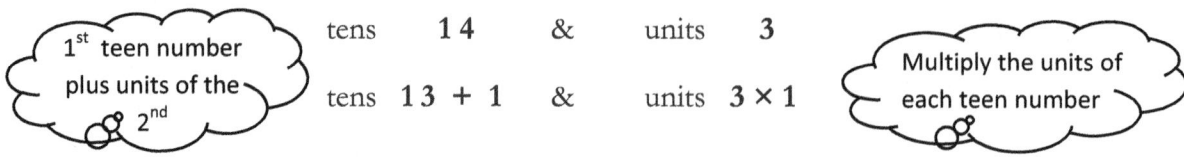

tens **14** & units **3**

tens **13 + 1** & units **3 × 1**

2) $\qquad 14 \times 12$

$\qquad$ tens $14 + 2 = \mathbf{16}$ & units $4 \times 2 = \mathbf{8}$

$\qquad\qquad \mathbf{14 \times 12 = 168}$

3) $\qquad 16 \times 11$

$\qquad$ tens $16 + 1 = \mathbf{17}$ & units $6 \times 1 = \mathbf{6}$

$\qquad\qquad \mathbf{16 \times 11 = 176}$

If the units product is greater or equal to 10, then we have to add the carry to the sum.

4) $\qquad 15 \times 13$

$\qquad$ tens $15 + 3 = \mathbf{18}$ & units $5 \times 3 = {}^1\mathbf{5}$

$\qquad\qquad 18 + 1 = \mathbf{19}$

$\qquad\qquad \mathbf{15 \times 13 = 195}$

5) $\qquad 19 \times 16$

$\qquad$ tens $19 + 6 = \mathbf{25}$ & units $9 \times 6 = {}^5\mathbf{4}$

$\qquad\qquad 25 + 5 = \mathbf{30}$

$\qquad\qquad \mathbf{19 \times 16 = 304}$

Practise questions

1) 13 × 14 2) 16 × 15

3) 19 × 17 4) 15 × 18

5) 13 × 16 6) 16 × 18

7) 19 × 14 8) 18 × 17

Answers:

1) 182 2) 240 3) 323 4) 270
5) 208 6) 288 7) 266 8) 306

Multiplying two numbers that lie between 30 & 70.

This trick uses a similar technique to multiplying numbers near 100, as fifty is ½ Base 100.

The shortcut if both numbers are in the fifties.

1)   5 4 × 5 6

   i.   Take the difference of each number from 50, and write this difference immediately above the number in a circle.

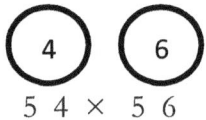

   ii.   Multiplying these two circled numbers.

$4 \times 6 = 24$

		2	4

   iii.   As the numbers are above 50, we cross add and halve that sum {since 50 is half 100}

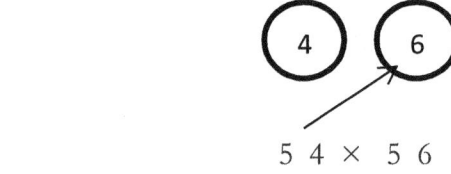

½ (54 + 6)
½ 60
= 30

3	0	2	4

$4 \times 6 = 24$

5 4 × 5 6 = 3,024

2)   5 3 × 5 9

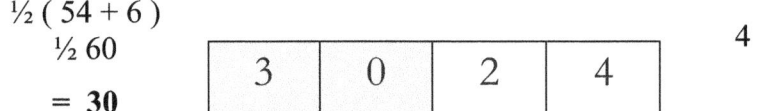

		2	7

$3 \times 9 = 27$

we cross add and halve

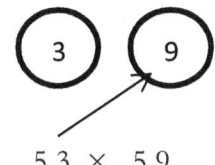

53 × 59

½ (53 + 9)
½ 62
= **31**

| 3 | 1 | 2 | 7 |

3 × 9 =
27

5 3 × 5 9 = 3,1 2 7

3) 5 8 × 5 5 =

5 8 × 5 5

| | | 4 | 0 |

8 × 5
40

we cross add and halve

31.5 is really 31 hundreds and 5 tens. So we add 5 tens to our tens digit.

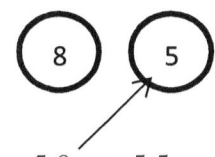

5 8 × 5 5
5

½ (58 + 5)
½ 63
= **31.5**

| 3 | 1 | 9 | 0 |

8 × 5
40
+ 50
90

5 8 × 5 5 = 3,1 9 0

We can use the same method for any numbers between 30 & 70. We need to be aware of 2 cases

1. If the two numbers are on either side of 50, this will require a trade.
2. If the product of the two differences is more than 100, it will involve a *carry*

Two numbers above 50

4) $\quad 62 \times 54 =$

$\quad\quad\quad\quad\quad\quad\quad\quad\quad\quad$ (12) (4)
$\quad\quad\quad\quad\quad\quad\quad\quad\quad\quad$ 6 2 × 5 4

½ (62 + 4) $\quad$ | 3 | 3 | 4 | 8 | $\quad$ 12 × 4
33 $\quad\quad\quad\quad\quad\quad\quad\quad\quad\quad\quad\quad\quad\quad\quad\quad$ **48**

$\quad\quad\quad\quad\quad\quad$ 6 2 × 5 4 = 3, 3 4 8

5) $\quad 68 \times 64 =$

$\quad\quad\quad\quad\quad\quad\quad\quad\quad\quad$ (18) (14)
$\quad\quad\quad\quad\quad\quad\quad\quad\quad\quad$ 6 8 × 6 4

$\quad\quad\quad\quad\quad\quad\quad\quad$ 2
½ (68 + 14) + 2 $\quad$ | 4 | 3 | 5 | 2 | $\quad$ 18 × 14
½ 82 + 2 $\quad\quad\quad\quad\quad\quad\quad\quad\quad\quad\quad\quad\quad\quad$ ² **52**
= **43**

$\quad\quad\quad\quad\quad\quad$ 6 8 × 6 4 = 4, 3 5 2

Two numbers below 50

6) $\quad 48 \times 43 =$

$\quad\quad\quad\quad\quad\quad\quad\quad\quad\quad$ (-2) (-7)
$\quad\quad\quad\quad\quad\quad\quad\quad\quad\quad$ 4 8 × 4 3

$\quad\quad\quad\quad\quad\quad\quad\quad$ 5
½ (48 – 7) $\quad\quad$ | 2 | 0 | 6 | 4 | $\quad$ -2 × -7
½ 41 $\quad\quad\quad\quad\quad\quad\quad\quad\quad\quad\quad\quad\quad\quad\quad$ = 14
= **20.5** $\quad\quad\quad\quad\quad\quad\quad\quad\quad\quad\quad\quad\quad\quad$ + 50
$\quad\quad\quad\quad\quad\quad\quad\quad\quad\quad\quad\quad\quad\quad\quad\quad\quad\quad$ **64**
Remember 20.5 is $\quad$ 4 8 × 4 3 = 2, 0 6 4
20 hundreds & **5** tens

7) $40 \times 37 =$

 ⓘ−10 ⓘ−13
 40×37

½ (40 − 13) + 1 =
 14.5

	¹	5	
1	4	8	0

−10 × −13
= ¹30
 + 50
 ¹ 80

Remember 14.5 is 14 hundreds & 5 tens

$40 \times 37 = 1,480$

Numbers on either side of 50

Remember the product of a positive number and a negative numbers is always negative and will require a trade.

8) $54 \times 48 =$

 ⓘ4 ⓘ−2 4× −2
 54×48 = −8

½ (54 − 2) − 1
 26 − 1
 = 25

2	5	9	2

We trade 100
100 − 8 = **92**

$54 \times 48 = 2,592$

9) $58 \times 46 =$

 ⓘ8 ⓘ−4 8× −4
 58×46 = −32

½ (58 − 4) − 1
 27 − 1
 = 26

2	6	6	8

We trade 100
100 − 32 = **68**

$58 \times 46 = 2,668$

10) $66 \times 31 =$

$$\overset{\displaystyle \boxed{16} \boxed{-19}}{66 \times 31}$$

$16 \times -19 = -304$

½ (66 – 19) – 4 + 1
23.5 – 3
= **20.5**

2	¹0	⁵4	6

$66 \times 31 = 2,046$

We trade 400
400 – 304
= 96
+ 50
¹ **46**

Practise questions

1) 56×52 2) 55×54

3) 69×57 4) 65×68

5) 41×47 6) 48×41

7) 37×39 8) 38×35

9) 44×56 10) 58×41

1) 2,912   2) 2,970   3) 3,933   4) 4,420
5) 1,927   6) 2,208   7) 1,806   8) 1,330
9) 2,464   10) 2,378

Two numbers in the same decade & the units add to 10

If the two numbers are in the same decade   20s, 30s, 40s … 90s and the units of the two numbers add to ten, the hundreds digit(s) is found by multiplying the tens digit by the next consecutive number. The last two digits are found by multiplying the units digits.

Let's look at $43 \times 47 = \mathbf{2,021}$ & $62 \times 68 = \mathbf{4,216}$

1) $43 \times 47 =$

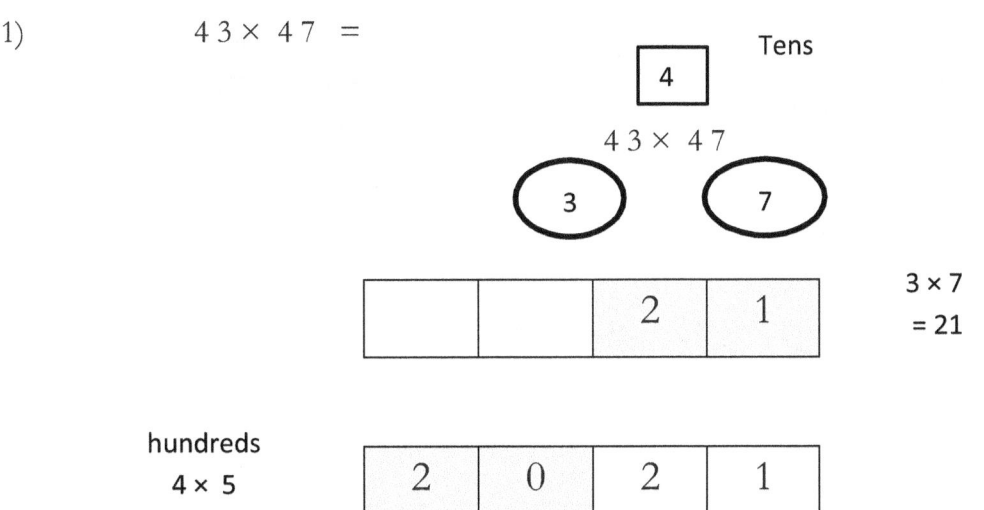

$43 \times 47 = 2,021$

2) $62 \times 68 =$

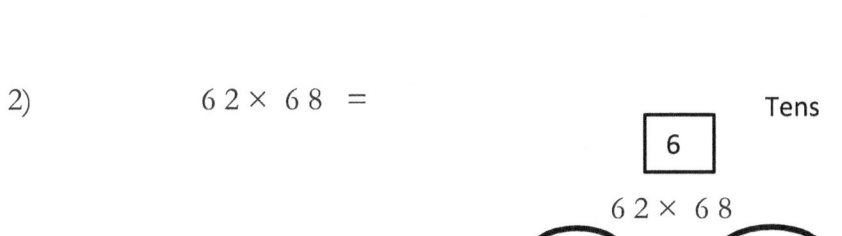

$62 \times 68 = 4,216$

3) 81×89

hundreds $(8 \times 9 = 72)$ & *units* $(1 \times 9 = 09)$

$$81 \times 89 = 7,209$$

4) 78×72

hundreds $(7 \times 8 = 56)$ & *units* $(8 \times 2 = 16)$

$$78 \times 72 = 5,616$$

5) 23×27

hundreds $(2 \times 3 = 6)$ & *units* $(3 \times 7 = 21)$

$$23 \times 27 = 621$$

Practise questions

1) 63×67 2) 86×84

3) 35×35 4) 78×72

5) 64×66 6) 57×53

7) 41×49 8) 88×82

Answers:

1) 4,221   2) 7,224   3) 1,225   4) 5,616

5) 4,224   6) 3,021   7) 2,009   8) 7,216

Two numbers above and below the same tens

Multiplying two numbers that are above and below the same tens can also be found with a little bit of manipulation.

1) 68 × 74

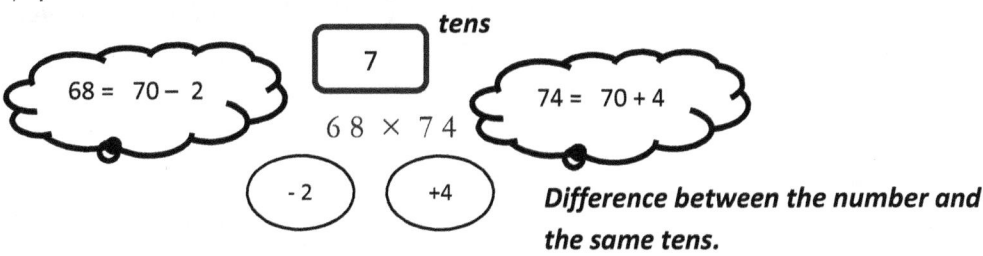

Difference between the number and the same tens.

Units
-2 × 4 = -8
Trade 10
10 − 8 = 2

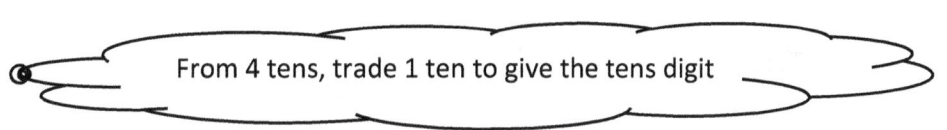

Tens digit   Add the two unit differences together and multiply by the tens digit

Tens   (-2 + 4) × 7 = 2 × 7 = 14

14 tens :− 1 hundred & 4 tens

From 4 tens, trade 1 ten to give the tens digit

Tens   4 − 1 = 3

| | | 3 | 2 |

Hundreds $7^2 + 1 = 50$

| 5 | 0 | 3 | 2 |

68 × 74 = 5,032

2)   7 6 × 8 5

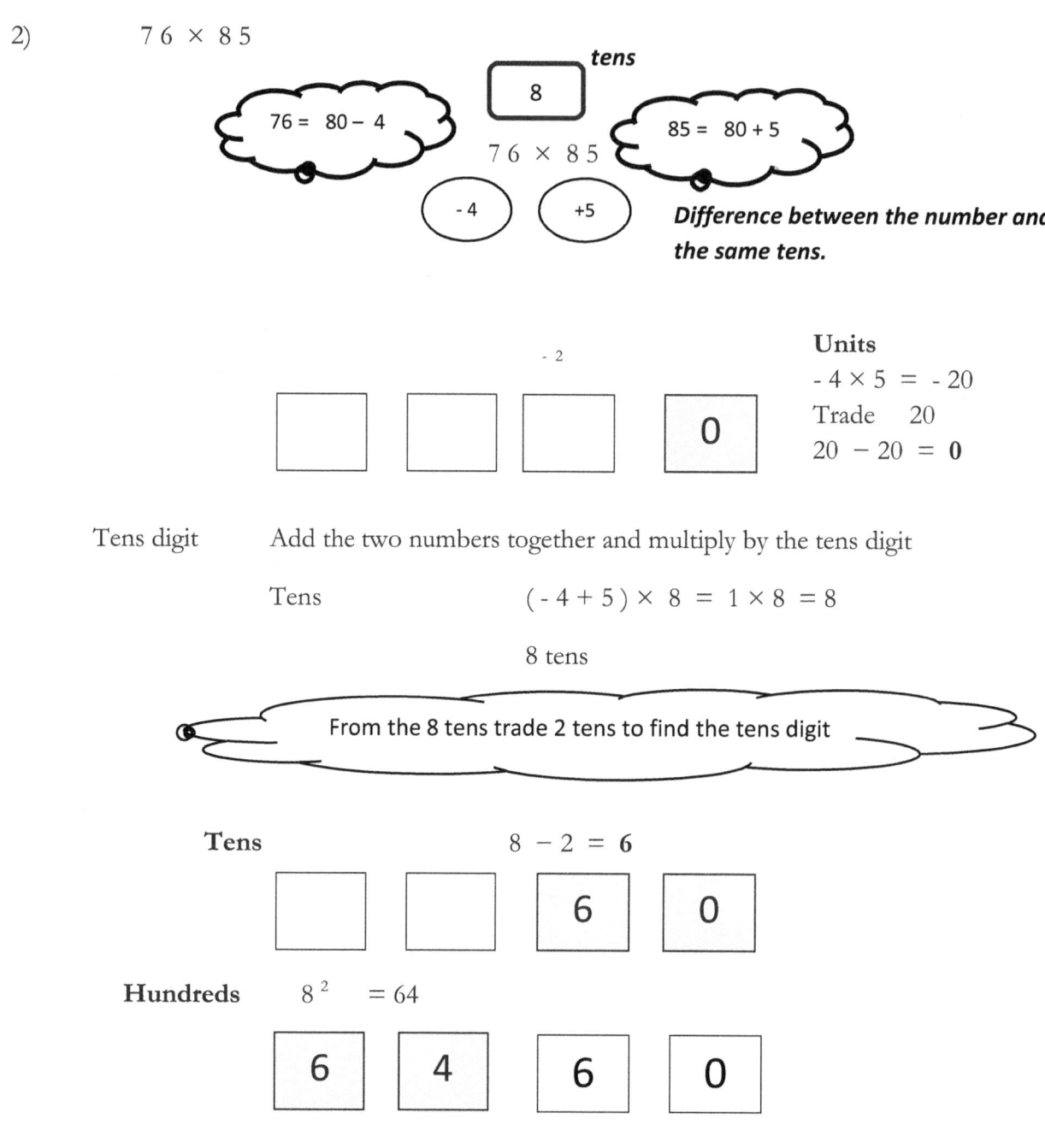

In the next example the both the product of units and product of tens are negative. This brings about a slight difference in the trading: {*The tens digit is found by adding the two totals together and trade from the upper base hundred*}.

3)   5 5 × 6 3

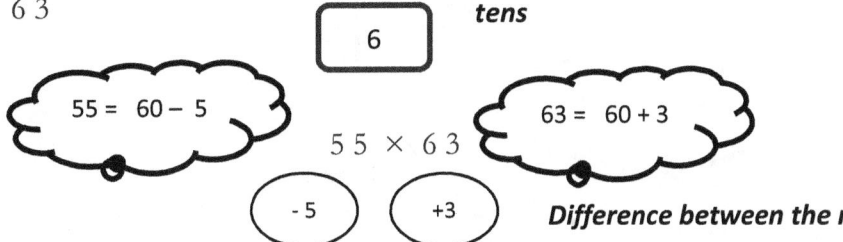

Difference between the number and the same tens.

           Units $-5 \times 3 = -15$

Tens digit   Add the two numbers together and multiply by the tens digit

           Tens $(-5 + 3) \times 6 = -2 \times 6 = -12$

                                  -12 tens

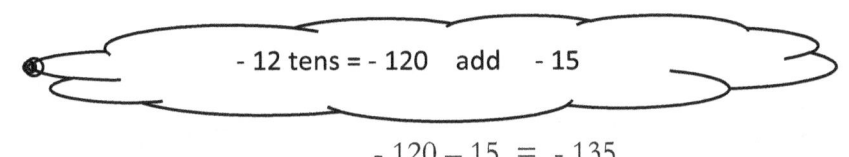

                          $-120 - 15 = -135$

We need to trade 2 hundreds, $200 - 135 = 65$

		6	5

Hundreds $6^2 - 2 \Leftrightarrow 36 - 2 = \mathbf{34}$

3	4	6	5

$5 5 \times 6 3 = 3{,}4 6 5$

4) 27×31

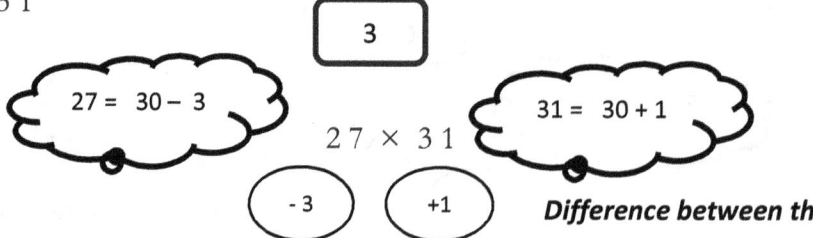

Difference between the number and the same tens.

Units $-3 \times 1 = -3$

Tens digit   Add the two numbers together and multiply by the tens digit

Tens $(-3 + 1) \times 3 = -2 \times 3 = -6$

             -6 tens

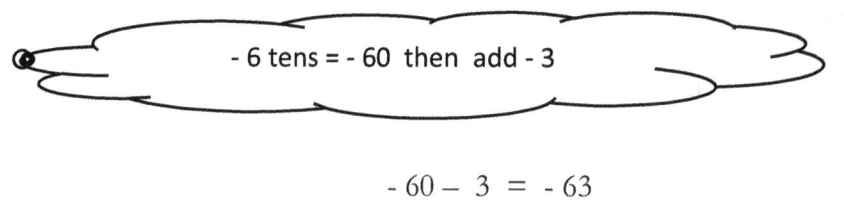

$-60 - 3 = -63$

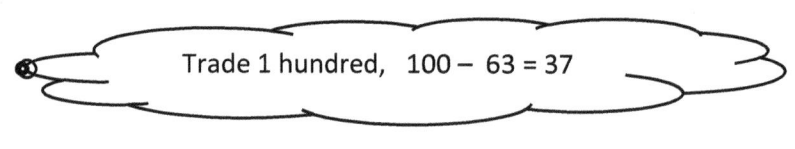

-1

	3	7

Hundreds $3^2 - 1 \Leftrightarrow 9 - 1 = 8$

8	3	7

$27 \times 31 = 837$

If the difference is the same, the method is even simpler.

5) 68 × 72

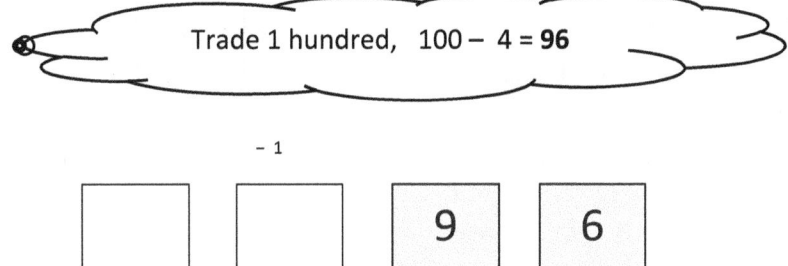

Units $-2 \times 2 = -4$

Tens digit   cancels out since $-2 + 2 = 0$

Trade 1 hundred, $100 - 4 = 96$

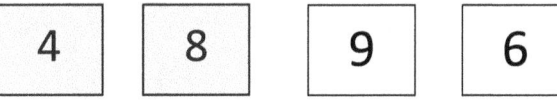

Hundreds $7^2 - 1 \Leftrightarrow 49 - 1 = 48$

| 4 | 8 | 9 | 6 |

68 × 72 = 4,896

6) 46 × 54

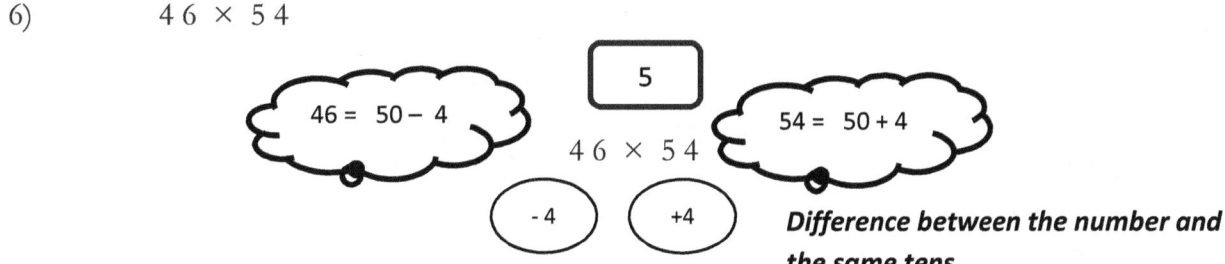

Units $\quad -4 \times 4 = -16$

Trade 1 hundred, $100 - 16 = \mathbf{84}$

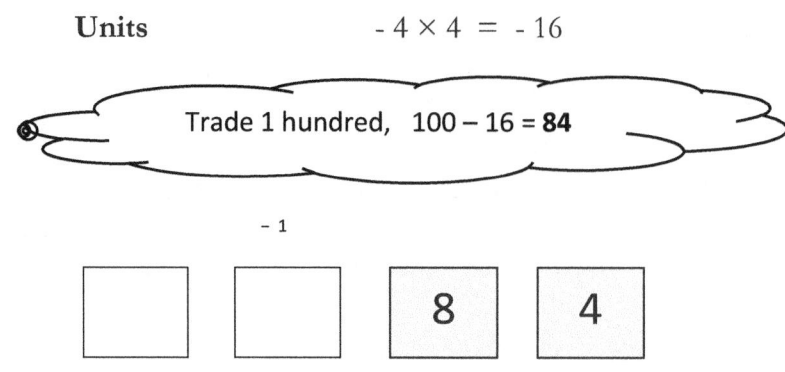

Hundreds $\quad 5^2 - 1 \Leftrightarrow 25 - 1 = \mathbf{24}$

$46 \times 54 = 2,484$

Practise questions

1) 73×65 2) 76×67

3) 85×79 4) 38×43

5) 67×72 6) 56×62

7) 48×52 8) 67×73

Answers:

1) 4,745 2) 5,092 3) 6,715 4) 1,634

5) 4,824 6) 3,472 7) 2,496 8) 4,891

Square numbers that end in 5

E.g. 15^2, 25^2, ... 95^2 and beyond

$15^2 = 225$ $25^2 = 625$ $35^2 = 1,225$ ....

Squaring a number that has a 5 as its unit digit, the answer always ends in twenty-five.

The hundreds digit(s) is the product of the ten(s) digit and the next consecutive number.

1) 35^2

   Hundreds | 1 | 2 | 2 | 5 | $5^2 = 25$
   $3 \times 4 = 12$

   $35^2 = 1,225$

2) 55^2

   hundreds $5 \times 6 = 30$ &   *units* $= 25$

   $55^2 = 3,025$

3) 95^2

   hundreds $9 \times 10 = 90$ &   *units* $= 25$

   $95^2 = 9,025$

This method still works for numbers greater than 100.

4) 115^2

   hundreds $11 \times 12 = 132$ &   *units* $= 25$

   $115^2 = 13,225$

5) 185^2

   hundreds $18 \times 19 = 342$ &   *units* $= 25$

   $185^2 = 34,225$

Practise questions

1) 15 × 15   2) 85²

3) 35 × 35   4) 45²

5) 65 × 65   6) 55²

7) 115 × 115   8) 125²

9) 145 × 145   10) 165²

Answers:

1) 225   2) 7,225   3) 1,225   4) 2,025
5) 4,225   6) 3,025   7) 13,225   8) 15,625
9) 21,025   10) 27,225

Square numbers close to 100

E.g. $98^2, 93^2, 88^2, 104^2, 109^2, 115^2$

This method uses the technique of numbers close to 100

Numbers below 100

1) 98^2

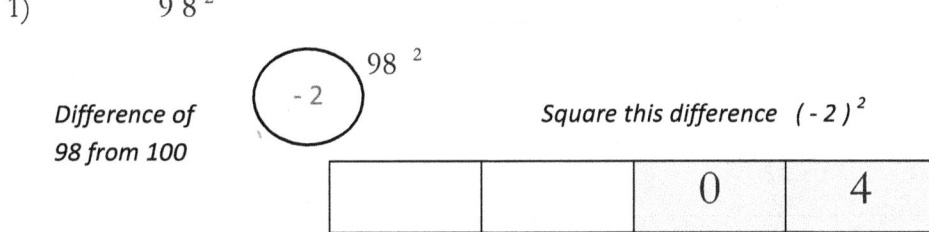

The number minus the difference   98 − 2

9	6	0	4

$98^2 = 9,604$

2) 93^2

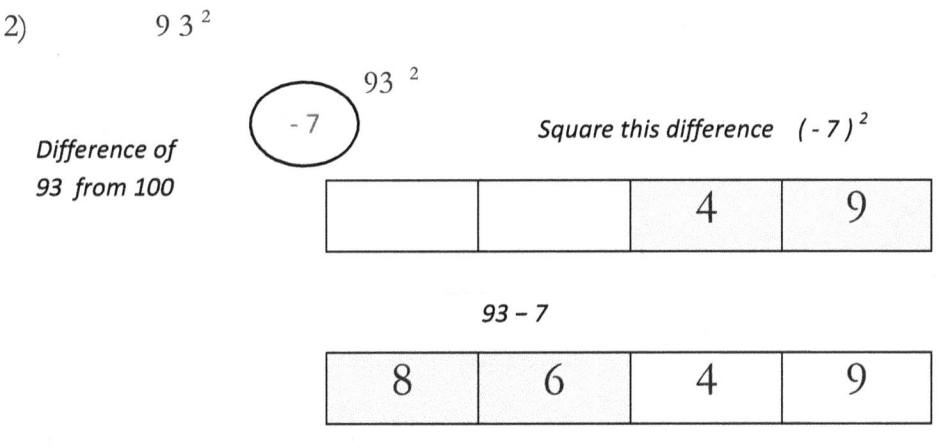

$93^2 = 8,649$

3) 88^2

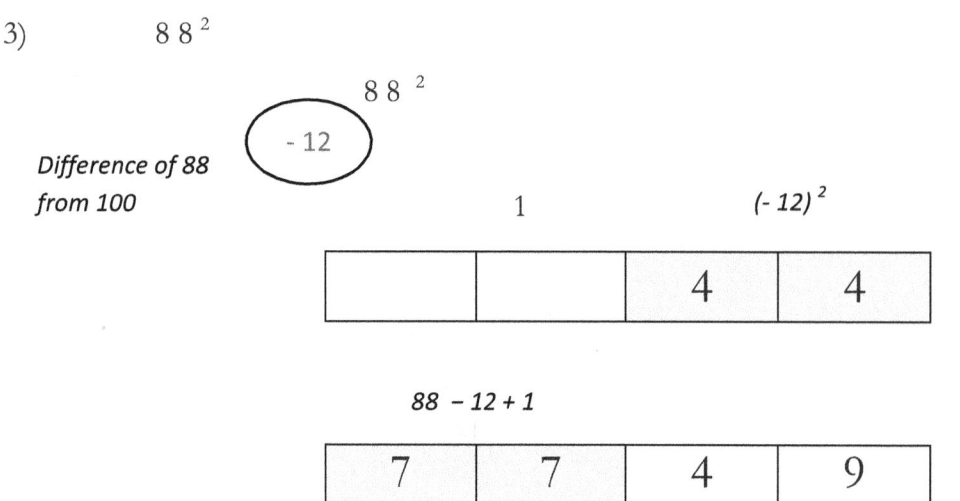

$88^2 = 7,744$

Numbers above 100

4) 104^2

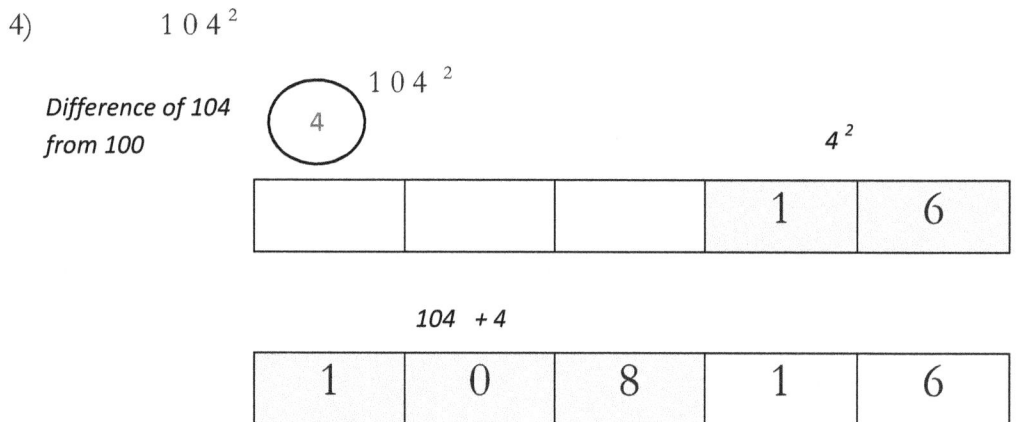

$104^2 = 10,816$

5) 109^2

Difference of 109 from 100 ⑨ 109^2 9^2

			8	1

$109 + 9$

1	1	8	8	1

$109^2 = 11,881$

6) 116^2

Difference of 116 from 100 ⑯ 116^2 16^2

2

			5	6

$116 + 16 + 2$

1	3	4	5	6

$116^2 = 13,456$

7) 108^2

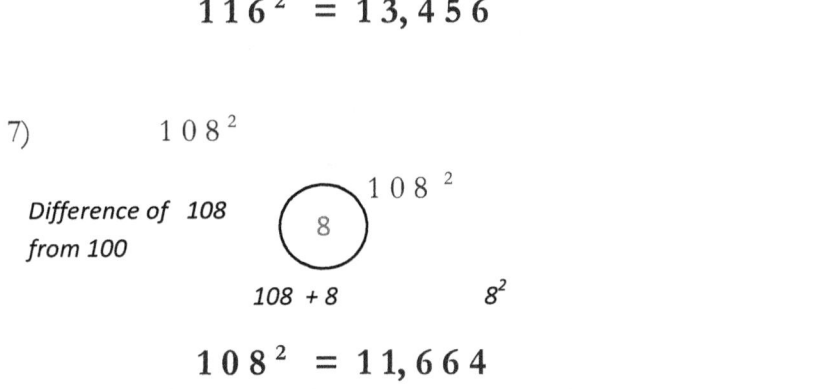

Difference of 108 from 100 ⑧ 108^2

$108 + 8 \qquad 8^2$

$108^2 = 11,664$

8) 94^2

Difference of 94 from 100   (-6) 94^2

    94 - 6 $(-6)^2$

$94^2 = 8,836$

Practise questions

1) 97×97 2) 91^2

3) 93×93 4) 92^2

5) 101×101 6) 103^2

7) 112×112 8) 114^2

Answers:

1) 9,409   2) 8,281   3) 8,649   4) 8,464

5) 12,201 6) 10,609 7) 12,544 8) 12,996

Squaring any number.

E.g. 13^2, 16^2, 24^2, 63^2, 79^2, 87^2 ..... and many more

To demonstrate the technique to squaring any 2 digit number, lets use 13^2, by first identifying the value of each digit.

$$13^2 \quad \text{breaking this up into tens and units}$$

$$(10 + 3)^2 \Leftrightarrow (10 + 3) \times (10 + 3)$$

$$10 \times 10 + 10 \times 3 + 10 \times 3 + 3 \times 3$$

$$10^2 + \text{twice}(10 \times 3) + 3^2$$

$$1^2 \text{ hundred} + 2 \times (1 \times 3) \text{ tens} + 3^2 \text{ units}$$

$$13^2 = 169$$

Examining each place value,

- o   the hundreds place is the **first digit squared**,
- o   the tens place is twice the **product** of the two digits
- o   the units place is the **second digit squared**.

When there is a **carry** we add it to the next place value.

Let's try this technique with the following example.

1) $\quad\quad\quad\quad\quad\quad\quad 16^2$

Hundreds	Tens	Units
$1^2 + 1 = 2$	$2 \times 1 \times 6 + 3$ $\quad ^15$	$6^2 = {}^36$
2	5	6

$$16^2 = 256$$

2) 24^2

Hundreds	Tens	Units
$2^2 + 1 = 5$	$2 \times 2 \times 4 + 1$	$4^2 = {}^1 6$
	${}^1 7$	

        5   7   6

$$24^2 = 576$$

3) 63^2

Hundreds	Tens	Units
$6^2 + 3 = 39$	$2 \times 6 \times 3$	$3^2 = 9$
	${}^3 6$	

    39   6   9

$$63^2 = 3,969$$

For squares where the number is near the upper tens, we can use the negative difference from the upper ten and trade from the hundreds digits.

4) [8] 79^2 (-1)

Hundreds	Tens	Units
$8^2 - 2 = 62$	$2 \times 8 \times -1$	$(-1)^2 = 1$
	-16	
	$20 - 16 = 4$	

    62   4   1

$$79^2 = 6,241$$

5) [9] 87^2 (-3)

Hundreds	Tens	Units
$9^2 - 6 = 75$	$2 \times 9 \times -3$	$(-3)^2 = 9$
	-54	
	$60 - 54 = 6$	

    75   6   9

$$87^2 = 7,569$$

Practise questions

1) 17×17 2) 14^2

3) 28×28 4) 32^2

5) 41×41 6) 53^2

7) 74×74 8) 82^2

9) 89×89 10) 67^2

11) 78×78 12) 91^2

Answers: 1) 289   2) 196   3) 784   4) 1,024
5) 1,681   6) 2,809   7) 5,476   8) 6,744
9) 7,921   10) 4,489   11) 6,084   12) 8,281

Multiplying by lots of 9s

The method for multiplying any number by the same quantity of 9s, (e.g. 47 × 99, or 234 × 999, or 7 456 × 9 999 or 31 946 × 99 999) is very simple and only requires using the 3rd method of subtraction.

- If the number of digits is more than the number of 9 digits, we have to do some simple manipulation to get our answer.
- If the number of digits is less than the number of 9s, it is even easier.

The key to this trick is knowing that 99 is 100 – 1, & 999 is 1 000 – 1, & 9 999 is 10 000 – 1, and so on.

Looking at the multiplication of 47×99 = $47 \times (100 - 1)$

= $4,700 - 47$

= $4,653$

Notice the tens and units digits 4 6 **5 3** are the difference between the multiplier and the upper base number (e.g. the difference between 47 & 100).

The first two digits **4 6** 5 3 is one less than the multiplier.

1) 85×99

The difference between the multiplier and 100

| | | 1 | 5 |

100 – 85

15

The one less than the multiplier

85 – 1

84

| 8 | 4 | 1 | 5 |

$85 \times 99 = 8,415$

2) 34×99

34 – 1   100 – 34

33   66

$34 \times 99 = 3,366$

Multiplying 3 digit numbers by 999   5 6 3 × 9 9 9

= 5 6 2, 4 3 7

The last three digits 5 6 2 **4 3 7** is the difference between the multiplier and the upper base number (e.g. the difference between 437 & 1 000)

The first three digits **5 6 2** 4 3 7 is one less than the multiplier.

3)   4 2 7 × 9 9 9

The difference between the multiplier and 1 000

| | | | 5 | 7 | 3 |

1 000 – 427

573

The one less than the multiplier

427 – 1

426

| 4 | 2 | 6 | 5 | 7 | 3 |

$$4\,2\,7 \times 9\,9\,9 \;=\; 4\,2\,6,5\,7\,3$$

4)   8 7 1 × 9 9 9

871 – 1   1, 000 – 871

870   **129**

$$8\,7\,1 \times 9\,9\,9 \;=\; 8\,7\,0,1\,2\,9$$

5)   4, 1 9 2 × 9, 9 9 9

4, 192 – 1   10, 000 – 4, 192

4, 191   **5, 808**

$$4,1\,9\,2 \times 9,9\,9\,9 \;=\; 4\,1,9\,1\,5,8\,0\,8$$

5) $\qquad 9,321 \times 9,999$

$\qquad\qquad 9,321 - 1 \qquad\qquad 10,000 - 9,321$

$\qquad\qquad\quad$ **9 320** $\qquad\qquad\qquad$ **0 6 7 9**

$\qquad\qquad\qquad\qquad$ *We must include the zero*

$$9,321 \times 9,999 = 93,200,679$$

The simplicity of this 'trick' will only work when the number of 9s is the same as the number of digits in the multiplier.

Practise questions

1) $\quad 17 \times 99$ $\qquad\qquad\qquad$ 2) $\quad 26 \times 99$

3) $\quad 72 \times 99$ $\qquad\qquad\qquad$ 4) $\quad 56 \times 99$

5) $\quad 341 \times 999$ $\qquad\qquad\qquad$ 6) $\quad 923 \times 999$

7) $\quad 5,528 \times 9,999$ $\qquad\qquad$ 8) $\quad 6,491 \times 9,999$

9) $\quad 27,833 \times 99,999$ $\qquad\;$ 10) $\quad 632,940 \times 999,999$

Answers:
1) 1,683   2) 2,574   3) 3,366   4) 5,544
5) 340,659   6) 426,573   7) 55,274,472   8) 64,903,509
9) 2,783,272,167   10) 632,939,367,060

Other cases

The number of 9s is less than the number of digits.

There are two different techniques depending whether the first digit(s) is smaller than larger than the last digit(s).

(1) $528 \times 99 = 528 \times (100 - 1)$

$52\,800 - 528$

$\quad 52\,800$
$\underline{-\quad 528}$

(2) $825 \times 99 = 825 \times (100 - 1)$

$82\,500 - 825$

$\quad 82\,500$
$\underline{-\quad 825}$

In each example, since we are multiplying by 99 (two 9s), the last two digits of the multiplier, will generate the tens and unit digits of the answer by finding the difference with the upper base number, (100).

In (1), if the first digit of the multiplier (**5**) is smaller than the last digit (**8**), trading will only affect the last digit of the multiplier, while the first digit(s) remains unchanged.

In (2), if the first digit of the multiplier (**8**) is (equal to or) larger than the last digit (**5**), the trading will affect the 2nd digit of the multiplier by the first digit(s) being reduced by 1.

1) $\qquad 528 \times 99$

We trade a 100 and subtract the last two digits of the multiplier from 100.

			7	2

100 − 28

72

Since the first digit of the multiplier '5' is less than the last digit of the multiplier '8', we reduce the last digit (8) by 1 (for the trade) and then subtract the '5'.

		2	7	2

7 − 5

2

The first two digits of the multiplier remain unchanged

5	2	2	7	2

$$528 \times 99 = 52{,}272$$

2) 825×99

| | | | 7 | 5 |

100 − 25

75

This is the 'tricky bit'.
We reduce the last digit '5' by one to get a working digit of '4'., But the first digit '8' is greater than the '4', we require a trade 'ten' from the 2nd digit (2). Thus you either build up from the '8' to the '4 - *teen*' or subtract (14 − 8).

| | | 6 | 7 | 5 |

14− 8

6

Because we built up or traded, we reduce the 2nd digit by 1. The first digit remains unchanged.

| 8 | 1 | 6 | 7 | 5 |

82− 1

81

$$825 \times 99 = 81,675$$

3) 715×99

| | | | 8 | 5 |

100 − 15

85

We reduce the last digit '5' by one to get a working number of '4'. But the first digit '7' is greater than '4', so we build up from the '7' to the '4 - *teen*'. This means we are finding the difference between 14 & 7, thus requiring a trade.

| | | 7 | 8 | 5 |

14− 7

7

Because we built up (traded), 71 is reduced by 1, to give 70.

| 7 | 0 | 7 | 8 | 5 |

71− 1

70

$$715 \times 99 = 70,785$$

4) $\qquad 2,847 \times 99$

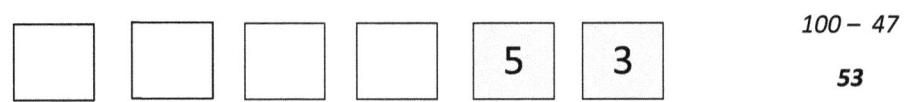

100 − 47

53

Because there are two extra digits, the '47' is reduced by 1 to become the working number of '46'. Since '46' is larger than the first two digits '28', we take the difference between these two numbers.

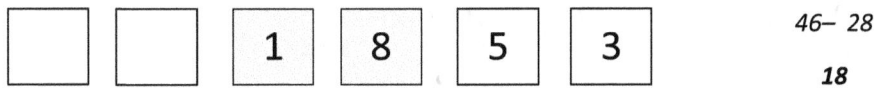

46− 28

18

There was no need to trade, thus we write the first two digits unchanged.

2	8	1	8	5	3

$$2,847 \times 99 = 281,853$$

5) $\qquad 8,562 \times 99$

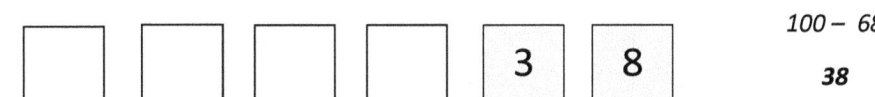

100 − 68

38

Because there are two extra digits, the last two digits '62' are reduced by 1 to become the working number of '61'. Since '61' is smaller than the first two digits '85', we need to build up from 85 to 61, or trade a 100 to get 161. Therefore 161 − 85 equals 76.

		7	6	3	8

161− 85

76

Since we built up (or traded), we reduce the 85 by one

8	4	7	6	3	8

85− 1

84

$$8,562 \times 99 = 847,638$$

6)   3,276 × 999

| | | | | 7 | 2 | 4 |

1 000 − 276
724

Because there is one extra digit, the last digit '6' is reduced by 1 to become the working number of '5'. Since '5' is larger than the first digit '3', we subtract 5 − 3.

| | | | 2 | 7 | 2 | 4 |

5 − 3
2

There was no need to trade in the above step so we write the first 3 digits as is

| 3 | 2 | 7 | 2 | 7 | 2 | 4 |

3,276 × 999 = 3,272,724

7)   58,125 × 999

| | | | | | 8 | 7 | 5 |

1 000 − 125
875

Because there are two extra digits, the last two digits '25' are reduced by 1 to become the working number of '24'. Since '24' is smaller than the first two digits '58', we need to build up from 58 to 24, or trade 100 to become 124 − 58 to equal 66.

| | | | 6 | 6 | 8 | 7 | 5 |

124 − 58
66

Since we traded we reduce the number 581 by one.

| 5 | 8 | 0 | 6 | 6 | 8 | 7 | 5 |

581 − 1
580

58,125 × 999 = 58,066,875

Practise questions

1) 17 × 9				2) 26 × 9

3) 38 × 9				4) 52 × 9

5) 84 × 9				6) 545 × 99

7) 953 × 99				8) 791 × 99

9) 7,866 × 99			10) 2,978 × 99

11) 2,368 × 999			12) 9,435 × 999

Answers:

1) 153   2) 234   3) 342   4) 468
5) 756   6) 53,955   7) 94,347   8) 78,309
9) 778,734 10) 294,822 11) 2,365,632 12) 9,425,565

The number of 9s is more than the number of digits.

Looking at the multiplication of $\quad 8 \times 9\,9$

$\quad\quad\quad\quad\quad\quad\quad\quad\quad\quad\quad\quad 8 \times (100 - 1)$

$\quad\quad\quad\quad\quad\quad\quad\quad\quad\quad\quad\quad 8\,0\,0 - 8$

$\quad\quad\quad\quad\quad\quad\quad\quad\quad\quad\quad\quad 7\,9\,2$

The first digit is one less than the multiplier. The units digit is the multiplier's best friend. The number of 9s occupying the middle digits depends on how many more 9s digits there are in comparison to the multiplier. Since the multiplicand has one more 9, we insert a 9 as the middle digit.

1) $\quad\quad\quad\quad\quad\quad\quad\quad\quad\quad 7 \times 9\,9$

$\quad\quad\quad\quad\quad 7 - 1 \quad\quad\quad$ One extra 9 $\quad\quad 10 - 7$

$\quad\quad\quad\quad\quad\quad 6 \quad\quad\quad\quad\quad\quad 9 \quad\quad\quad\quad\quad\quad 3$

$\quad\quad 7 \times 9\,9 \,=\, 6\,9\,3$

2) $\quad\quad\quad\quad\quad\quad\quad\quad\quad\quad 3 \times 9\,9\,9$

$\quad\quad\quad\quad\quad 3 - 1 \quad\quad\quad$ Two extra 9s $\quad\quad 10 - 3$

$\quad\quad\quad\quad\quad\quad 2 \quad\quad\quad\quad\quad\quad\quad\quad\quad\quad\quad\quad 7$

$\quad\quad\quad\quad\quad\quad\quad\quad\quad\quad\quad\quad 9\,9$

$\quad\quad 3 \times 9\,9\,9 \,=\, 2,9\,9\,7$

3) $\quad\quad\quad\quad\quad\quad\quad\quad\quad\quad 5 \times 9,9\,9\,9$

$\quad\quad\quad\quad\quad 5 - 1 \quad\quad\quad$ Three extra 9s $\quad\quad 10 - 5$

$\quad\quad\quad\quad\quad\quad 4 \quad\quad\quad\quad\quad\quad\quad\quad\quad\quad\quad\quad 5$

$\quad\quad\quad\quad\quad\quad\quad\quad\quad\quad\quad\quad 9\,9\,9$

$\quad\quad 3 \times 9,9\,9\,9 \,=\, 4\,9,9\,9\,5$

4) $\quad\quad\quad\quad\quad\quad\quad\quad\quad\quad 8 \times 9\,9,9\,9\,9$

$\quad\quad\quad\quad\quad 8 - 1 \quad\quad\quad$ Four extra 9s $\quad\quad 10 - 8$

$\quad\quad\quad\quad\quad\quad 7 \quad\quad\quad\quad\quad\quad\quad\quad\quad\quad\quad\quad 2$

$\quad\quad\quad\quad\quad\quad\quad\quad\quad\quad\quad\quad 9\,9\,9\,9$

$\quad\quad 8 \times 9\,9,9\,9\,9 \,=\, 7\,9\,9,9\,9\,2$

5)
$$26 \times 999$$

26 − 1	One extra 9	100 − 26
25	**9**	**74**

$$26 \times 999 = 25{,}974$$

6)
$$64 \times 9{,}999$$

64 − 1	Two extra 9s	100 − 64
63	**99**	**36**

$$64 \times 9{,}999 = 639{,}936$$

7)
$$72 \times 99{,}999$$

72 − 1	Three extra 9s	100 − 72
71	**999**	**28**

$$72 \times 99{,}999 = 7{,}199{,}928$$

8)
$$829 \times 99{,}999$$

829 − 1	Two extra 9s	1\,000 − 829
828	**99**	**171**

$$829 \times 99{,}999 = 82{,}899{,}171$$

9)
$$4{,}138 \times 99{,}999$$

4, 138 − 1	One extra 9	10, 000 − 4, 138
4, 137	**9**	**5, 862**

$$4{,}138 \times 99{,}999 = 413{,}795{,}862$$

Practise questions

1)   8 × 9   2)   6 × 999

3)   43 × 999   4)   83 × 9,999

5)   94 × 99,999   6)   542 × 9,999

7)   953 × 9,999   8)   868 × 99,999

9)   4,269 × 99,999   10)   6,231 × 999,999

Answers:

1) 792   2) 5,994   3) 42,958   4) 829,917

5) 9,399,906   6) 5,419,458   7) 9,529,047

8) 86,799,132   9) 426,895,731   10) 6,230,993,769

"AREN'T THERE ENOUGH PROBLEMS IN THE WORLD ALREADY?"

Chapter 6

3 digit by 3 digit multiplication

If you master the skills presented in this chapter, then any fear of multiplying 3 digit numbers should be eased if not disappear altogether.

Remember visit the web site *modmaths.com.au*, to help refresh your memory of the various techniques.

Multiplication of three digit numbers.

The Vedic pattern of multiplying three digit numbers builds on the pattern of the two digits.

Using the technique of looking for the pattern, we let each digit be represented by a letter.

		a	b	c	×
		d	e	f	
	af	bf	cf		
	ae	be	ce	0	
ad	bd	cd	0	0	

The pattern of expansion for 3 digit by 3 digit multiplication uses a vertical and cross-wise connection of pairs of digits.

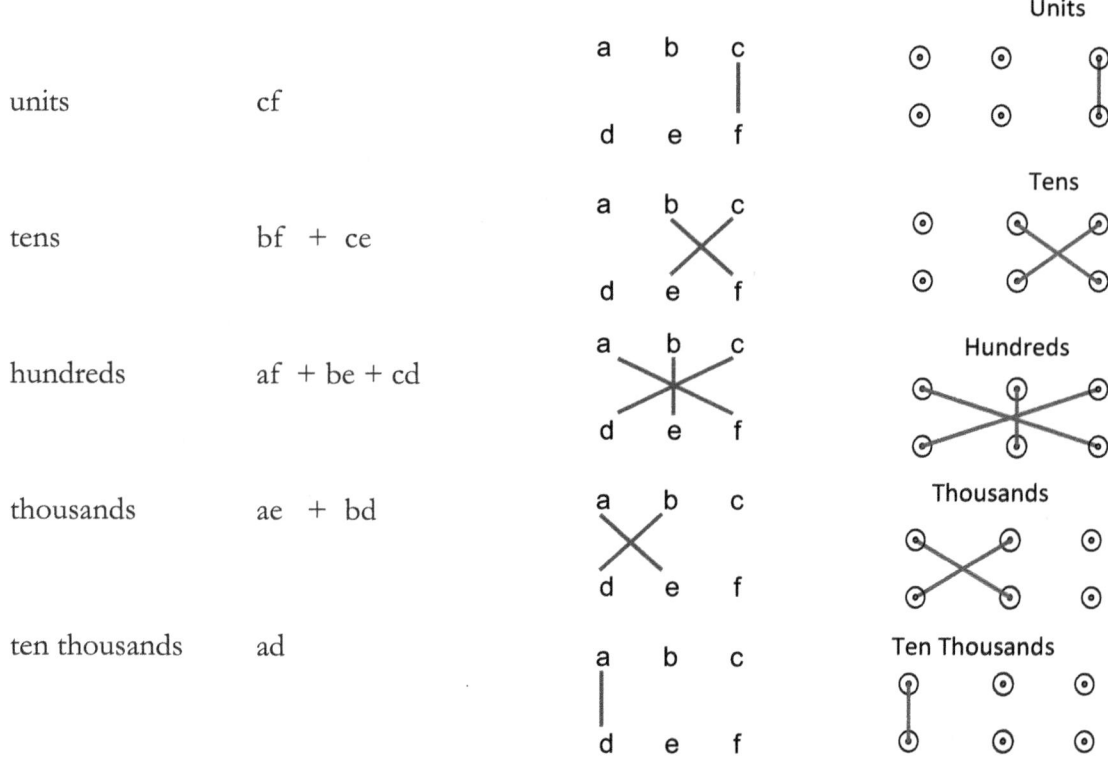

As you can see this pattern of multiplication is similar to the two-digit pattern and has a symmetry of calculations. This vertical and cross-wise pattern can be used by multiplying any number of digits. In the pattern there is a priority of connecting pairs of digits; working from the outside in, the cross-wise connection takes precedence before the vertical connection. This can be seen in the following 4 digit and 5 digit expansion.

The one difficulty of using this method to solve 3 digit and higher multiplications is that the middle term may require adding 3 or more large products. The pattern, however, is very important to master.

Let's begin with an easier problem to demonstrate the technique.

1) 123×241

Units	$1\ 2\ 3\ \times$	Units
$3 \times 1 = 3$	$2\ 4\ 1$	
	$\overline{3}$	

Tens	$1\ 2\ 3\ \times$	Tens
$2 \times 1 + 3 \times 4 = 14$	$2\ 4\ 1$	
4 & carry (1)	$\overline{^14\ 3}$	

Hundreds	$1\ 2\ 3\ \times$	Hundreds
$1 \times 1 + 2 \times 4 + 2 \times 3 + 1 = 16$	$2\ 4\ 1$	
6 & carry (1)	$\overline{^16\ ^14\ 3}$	

Thousands	$1\ 2\ 3\ \times$	Thousands
$1 \times 4 + 2 \times 2 = 9$	$2\ 4\ 1$	
	$\overline{9\ ^16\ ^14\ 3}$	

Ten thousands	$1\ 2\ 3\ \times$	Ten Thousands
$1 \times 2 = 2$	$2\ 4\ 1$	
	$\overline{2\ 9\ 6\ 4\ 3}$	

$$123 \times 241 = 29{,}643$$

2) $\qquad$ 3 2 4 × 4 1 2

Units $4 \times 2 = \mathbf{8}$	3 2 4 × 4 1 2 ――――― 8	Units
Tens $2 \times 2 + 4 \times 1 = \mathbf{8}$	3 2 4 × 4 1 2 ――――― 8 8	Tens
Hundreds $3 \times 2 + 2 \times 1 + 4 \times 4 = 24$ **4 & carry (2)**	3 2 4 × 4 1 2 ――――― ²4 8 8	Hundreds
Thousands $3 \times 1 + 2 \times 4 + 2 = 13$ **3 & carry (1)**	3 2 4 × 4 1 2 ――――― ¹3 ²4 8 8	Thousands
Ten thousands $3 \times 4 + 1 = \mathbf{13}$	3 2 4 × 4 1 2 ――――― 1 3 ¹3 ²4 8 8	Ten Thousands

$$3\,2\,4 \times 4\,1\,2 = 1\,3\,3,4\,8\,8$$

For larger 3 digit numbers the sum of the products can become difficult to calculate.

3) $\qquad 987 \times 898$

Units

$7 \times 8 = $ **56**

6 & carry (5)

$$\begin{array}{r} 9\ 8\ 7\ \times \\ 8\ 9\ 8 \\ \hline {}^5 6 \\ \hline \end{array}$$

Tens

$8 \times 8 + 9 \times 7 + 5 = $ **132**

2 & carry (13)

$$\begin{array}{r} 9\ 8\ 7\ \times \\ 8\ 9\ 8 \\ \hline {}^{13}2\ {}^5 6 \\ \hline \end{array}$$

Hundreds

$9 \times 8 + 8 \times 9 + 7 \times 8 + 13 = 213$

3 & carry (21)

$$\begin{array}{r} 9\ 8\ 7\ \times \\ 8\ 9\ 8 \\ \hline {}^{21}3\ {}^{13}2\ {}^5 6 \\ \hline \end{array}$$

Thousands

$9 \times 9 + 8 \times 8 + 21 = $ **166**

6 & carry (16)

$$\begin{array}{r} 9\ 8\ 7\ \times \\ 8\ 9\ 8 \\ \hline {}^{16}6\ {}^{21}3\ {}^{13}2\ {}^5 6 \\ \hline \end{array}$$

Ten thousands

$9 \times 8 + 16 = $ **88**

$$\begin{array}{r} 9\ 8\ 7\ \times \\ 8\ 9\ 8 \\ \hline 8 8\ {}^{16}6\ {}^{21}3\ {}^{13}2\ {}^5 6 \\ \hline \end{array}$$

$$987 \times 898 = 886{,}326$$

Short cuts to 3 digit by 3 digit multiplication

If the 3 digit numbers are near the "hundreds", either both above 605 × 708 or both below 496 × 891, we can apply a simpler method of multiplying the two numbers. This method is very similar to method used in the 2 digit calculations where the numbers were in the 90's or just above the 100.

If both numbers are greater than the 'hundreds'

1)   4 0 2 × 8 0 6

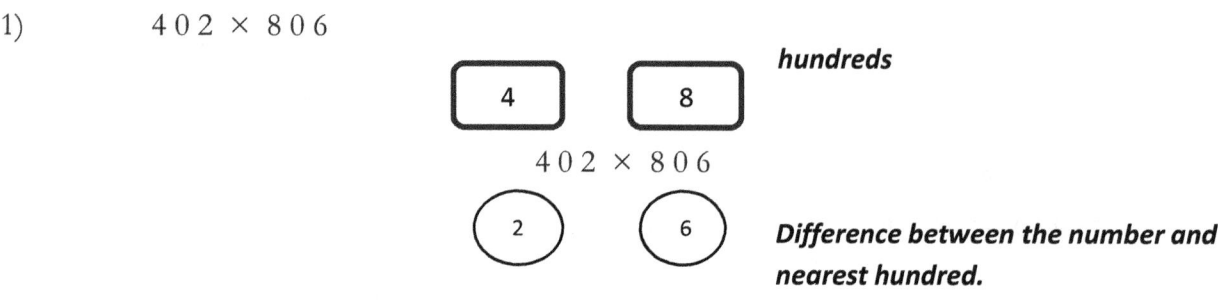

Most 3 digit by 3 digit multiplications will have 6 digit answers

The last two places are filled by the product of the difference digits 2 × 6 = 12

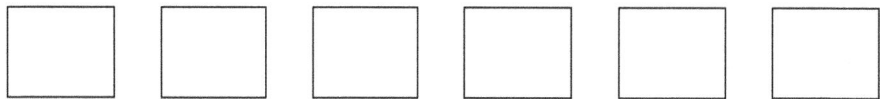

To find the middle two places we have to use the cross product

4 × 6 + 2 × 8 = 40

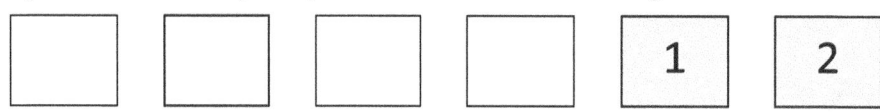

The first two places are filled by the product of the hundreds digits

4 × 8 = 32

| 3 | 2 | 4 | 0 | 1 | 2 |

4 0 2 × 8 0 6 = 3 2 4, 0 1 2

2)   712 × 304

               [7]   [3]   **hundreds**

            712 × 304

            (12)   (4)   *Difference between the number and nearest hundred.*

                                    12 × 4 = 48

    [ ] [ ] [ ] [ ] [4] [8]

The cross product   7 × 4 + 12 × 3 = 64, no carry

    [ ] [ ] [6] [4] [4] [8]

The first two places are filled by the product of the hundreds digits

    7 × 3 = 21

    [2] [1] [6] [4] [4] [8]

    712 × 304 = 216,448

If the two difference product is higher than 99, then we have to carry over the hundreds digit.

3)   8 1 2 × 7 1 1

   **hundreds**

8 1 2 × 7 1 1

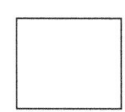 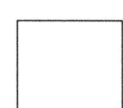   **Difference between the number and nearest hundred.**

12 × 11 = ^{1}32

				1	
				3	2

The cross product   8 × 11 + 12 × 7 = 172

172 + 1 = 173

= ^{1} 7 3   carry

		1		1	
		7	3	3	2

The first two places are filled by the product of the hundreds digits

8 × 7 + 1 = 56 + 1

= 57

		1		1	
5	7	7	3	3	2

8 1 2 × 7 1 1 = 5 7 7, 3 3 2

In the next example, the numbers are getting further away from the base number.

4)   416 × 817

```
    ┌───┐       ┌───┐      hundreds
    │ 4 │       │ 8 │
    └───┘       └───┘
         416 × 817
      ⎛ 16 ⎞    ⎛ 17 ⎞   Difference between the number and
      ⎝    ⎠    ⎝    ⎠   nearest hundred.
```

$$16 \times 17 = {}^2 72$$

```
                        2
┌──┐ ┌──┐ ┌──┐ ┌──┐ ┌──┐ ┌──┐
│  │ │  │ │  │ │  │ │ 7│ │ 2│
└──┘ └──┘ └──┘ └──┘ └──┘ └──┘
```

The cross product $4 \times 17 + 8 \times 16 = 196$ plus the carry

196 + the carry (2) *from 272*

196 + 2 = 198

= ^{1} 98

```
          1         2
┌──┐ ┌──┐ ┌──┐ ┌──┐ ┌──┐ ┌──┐
│  │ │  │ │ 9│ │ 8│ │ 7│ │ 2│
└──┘ └──┘ └──┘ └──┘ └──┘ └──┘
```

The first two places are filled by the product of the hundreds digits plus the carry

$4 \times 8 + 1 = 32 + 1$

$= 33$

```
     1         2
┌──┐ ┌──┐ ┌──┐ ┌──┐ ┌──┐ ┌──┐
│ 3│ │ 3│ │ 9│ │ 8│ │ 7│ │ 2│
└──┘ └──┘ └──┘ └──┘ └──┘ └──┘
```

416 × 817 = 339,872

With simplified working

5)   608 × 903

```
                    [6]        [9]
                    608 × 903
                    (8)        (3)
```

    6 × 9   (6 × 3 + 8 × 9)   8 × 3

     54   90   24

608 × 903 = 549,024

Practise questions

1)   501 × 212   2)   603 × 204

3)   704 × 903   4)   806 × 605

Answers:

1) 106, 212   2) 123, 012

3) 635, 712   4) 407, 630

132

6) {with carries}   6 0 9 × 8 1 2 =

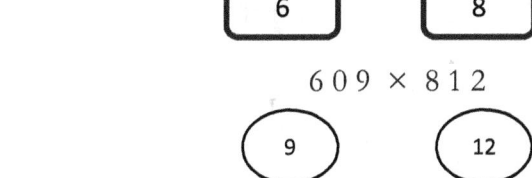

6 0 9 × 8 1 2

6 × 8	(9 × 8 + 6 × 12)	9 × 12
48	¹44	¹08
	¹ 44 + 1	08
48 + 1	**45**	**08**
49		

7 0 9 × 8 1 2 = 4 9 4, 5 0 8

Practise questions

5)   9 1 2 × 6 1 5   6)   8 1 4 × 7 1 1

7)   4 1 4 × 5 1 3   8)   3 0 6 × 7 2 0

Answers:
5)   560, 880   6)   578, 754

7)   212, 382   8)   220, 320

3 digit by 3 digit with both numbers less than the 'hundreds'

If both numbers are below the hundreds there has to be a slight adjustment to the above process to calculate the middle two digits.

If you remember when we had the two numbers in the 90's we had to subtract. In the following examples while we will still subtract, I will apply the 3rd method of subtraction, where we will calculate the difference between the next hundreds, and use these two digits in our answer. The first 2 digits will be reduced by the value of the hundreds.

1) $\qquad 494 \times 898 =$

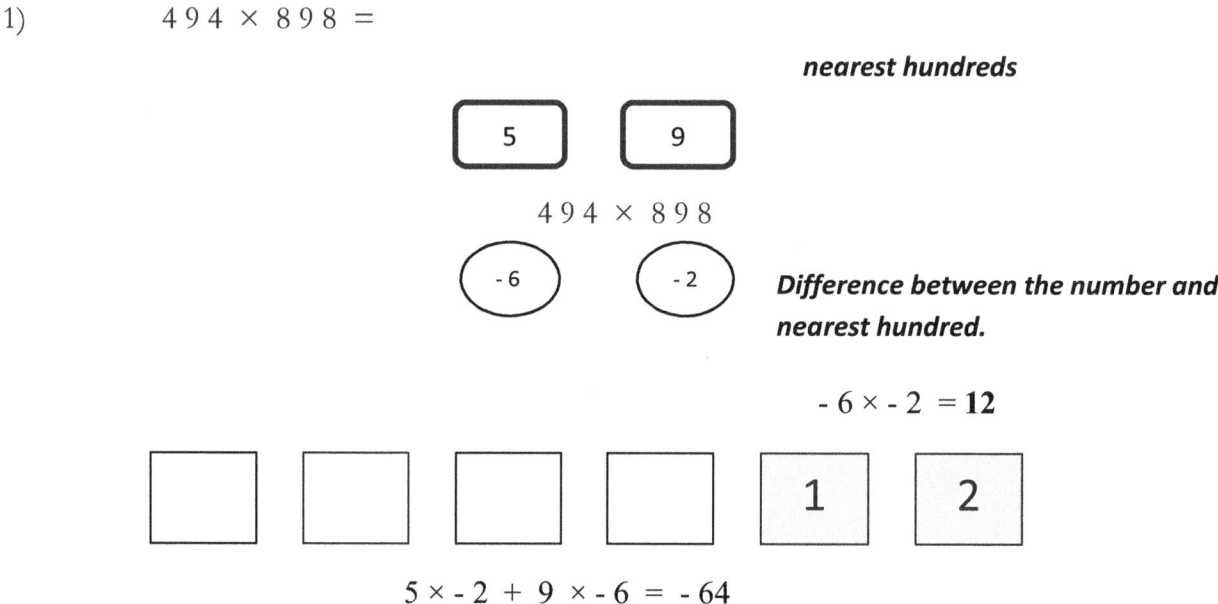

We take the difference between the number (64) and the upper hundred (100).

$$494 \times 898 = 443,612$$

2)   3 9 2 × 6 9 4

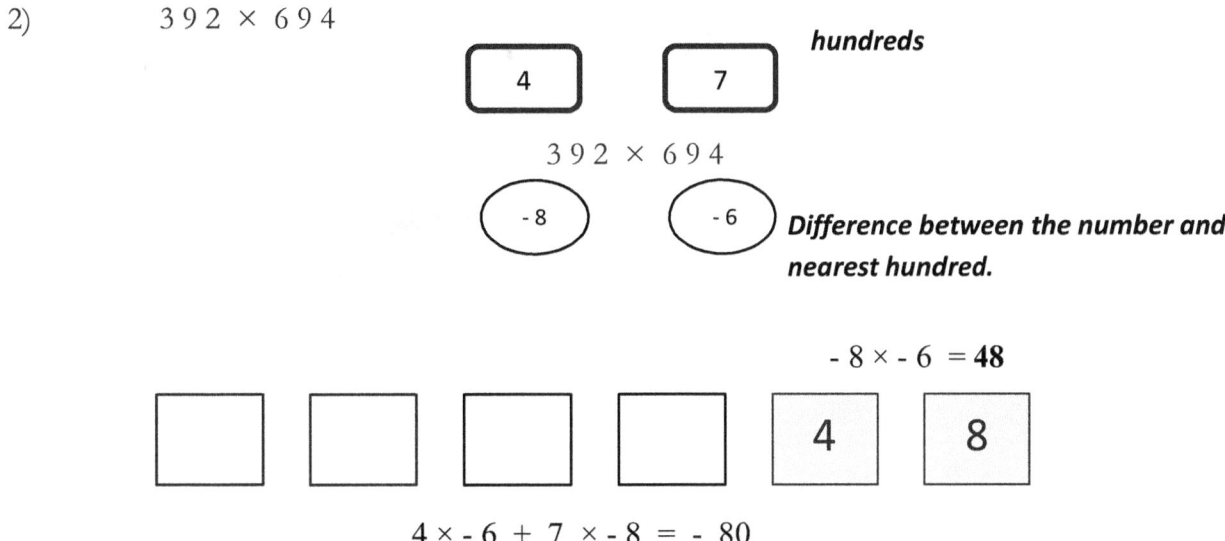

$-8 \times -6 = 48$

$4 \times -6 + 7 \times -8 = -80$

We take the difference between the number (80) and the upper hundred (100).

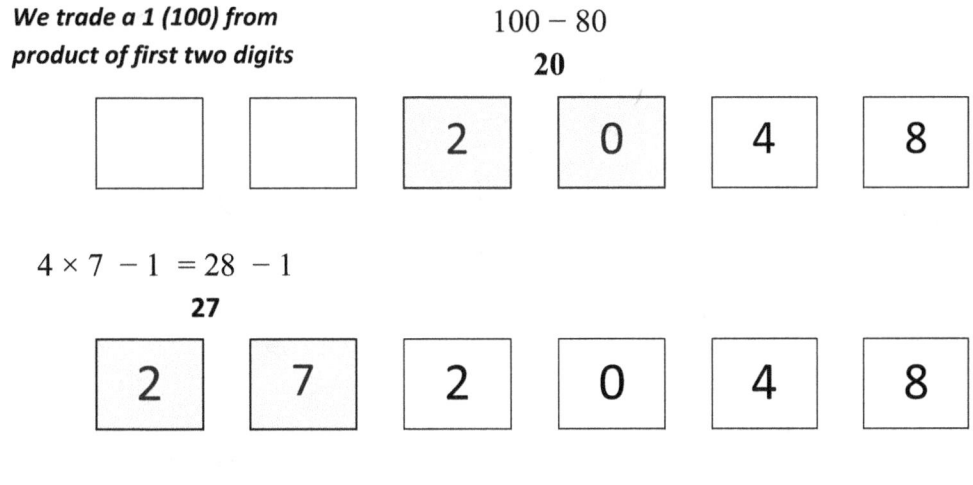

$4 \times 7 - 1 = 28 - 1$
        27

| 2 | 7 | 2 | 0 | 4 | 8 |

3 9 2 × 6 9 4 = 2 7 2 0 4 8

If the product of the two differences is higher than 99, then we have to adjust the trading process to the upper base hundred.

3)   788 × 489

[8]   [5]   **hundreds**

788 × 489

(-12)   (-11)   ***Difference between the number and nearest hundred.***

$-12 \times -11 = {}^1 32$

[ ] [ ] [ ] [¹] [3] [2]

$8 \times -11 + 5 \times -12 + 1 = -148 + 1$

We take the difference between the number (148) and the upper hundred (200).

$$200 - 148 + 1$$
$$= 52 + 1$$
$$53$$

We trade a 2 (200) from product of first two digits

[ ] [ ] [5] [3] [3] [2]

$8 \times 5 - 2 = 40 - 2$
38

[3] [8] [5] [3] [3] [2]

788 × 489 = 385,332

136

4)   6 8 6 × 7 8 4

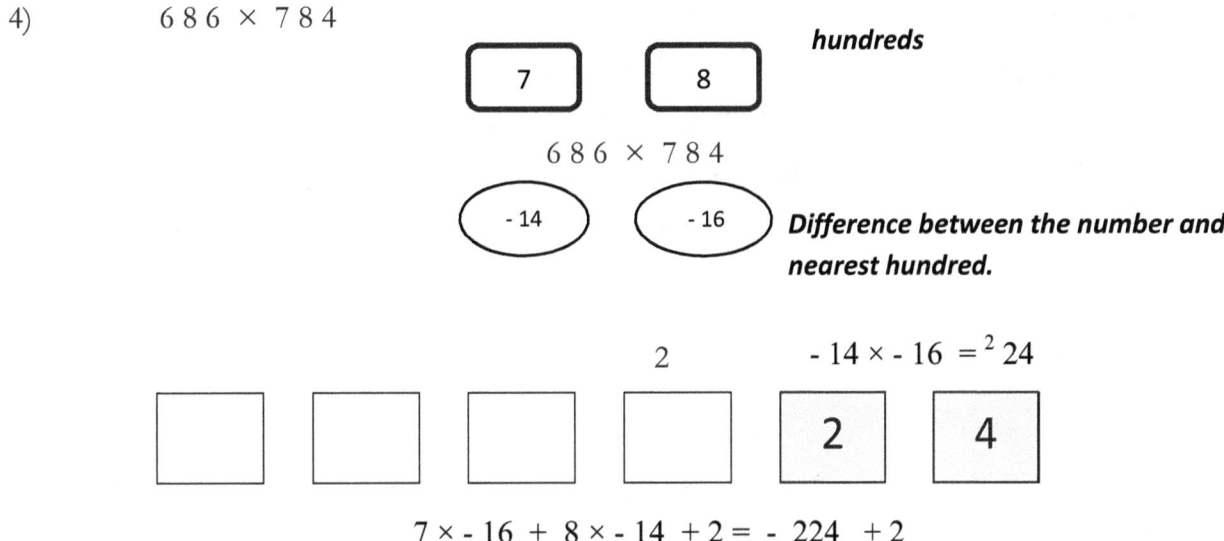

hundreds

Difference between the number and nearest hundred.

$-14 \times -16 = {}^224$

$7 \times -16 + 8 \times -14 + 2 = -224 + 2$

We take the difference between the negative number (224) and the immediately higher hundred (300).

We trade a 3 (300) from product of first two digits

$300 - 224 + 2$
$= 76 + 2$
$ 78$

$7 \times 8 - 3 = 56 - 3$

53

| 5 | 3 | 7 | 8 | 2 | 4 |

6 8 6 × 7 8 4 = 5 3 7, 8 2 4

As you can see from this example, when the differences get further and further away from the hundreds, the degree of difficulty in the multiplication increases.

With simplified working

5)   596 × 992

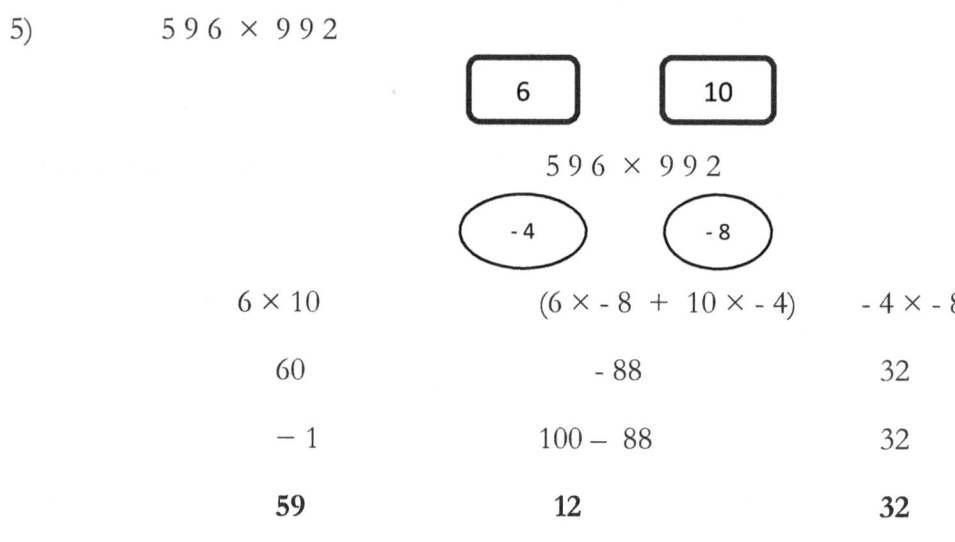

6 × 10	(6 × -8 + 10 × -4)	-4 × -8
60	-88	32
− 1	100 − 88	32
59	**12**	**32**

596 × 993 = 591,232

6) {with carries}   692 × 684 =

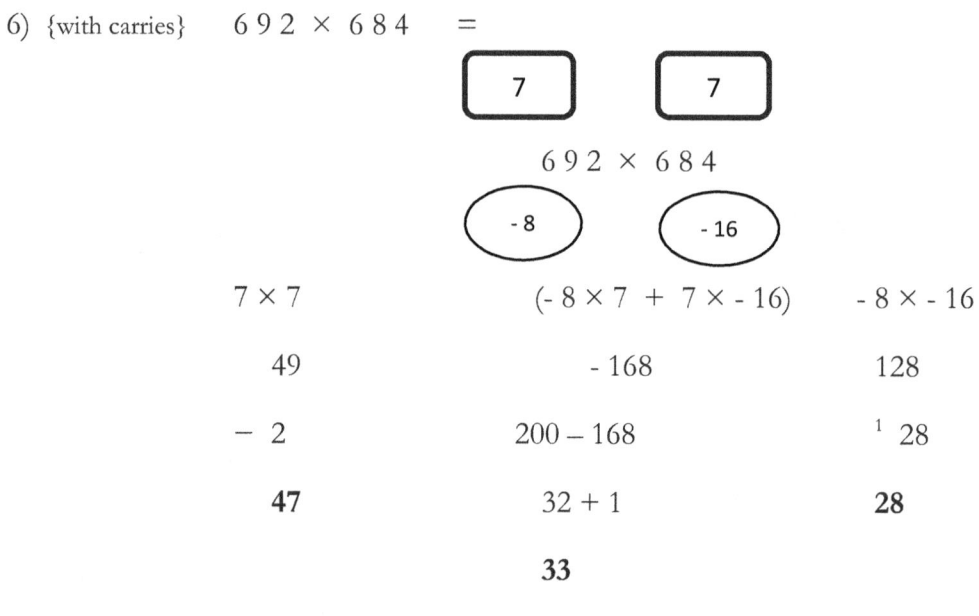

7 × 7	(-8 × 7 + 7 × -16)	-8 × -16
49	-168	128
− 2	200 − 168	¹ 28
47	32 + 1	**28**
	33	

692 × 684 = 473,328

Practise questions

1)   492 × 593   2)   693 × 295

Answers:

1)   291,756   2)   204,435

The product of numbers on either side of the hundreds.

This requires the skill of working with negative numbers.
- We will need to trade in the first step.
- In working out the sum of the second step, always begin with the positive terms and then subtract the second product.
- If the answer is a positive you will not need to trade in the 3rd step. If it is a negative you will need to trade.

1)   796 × 705

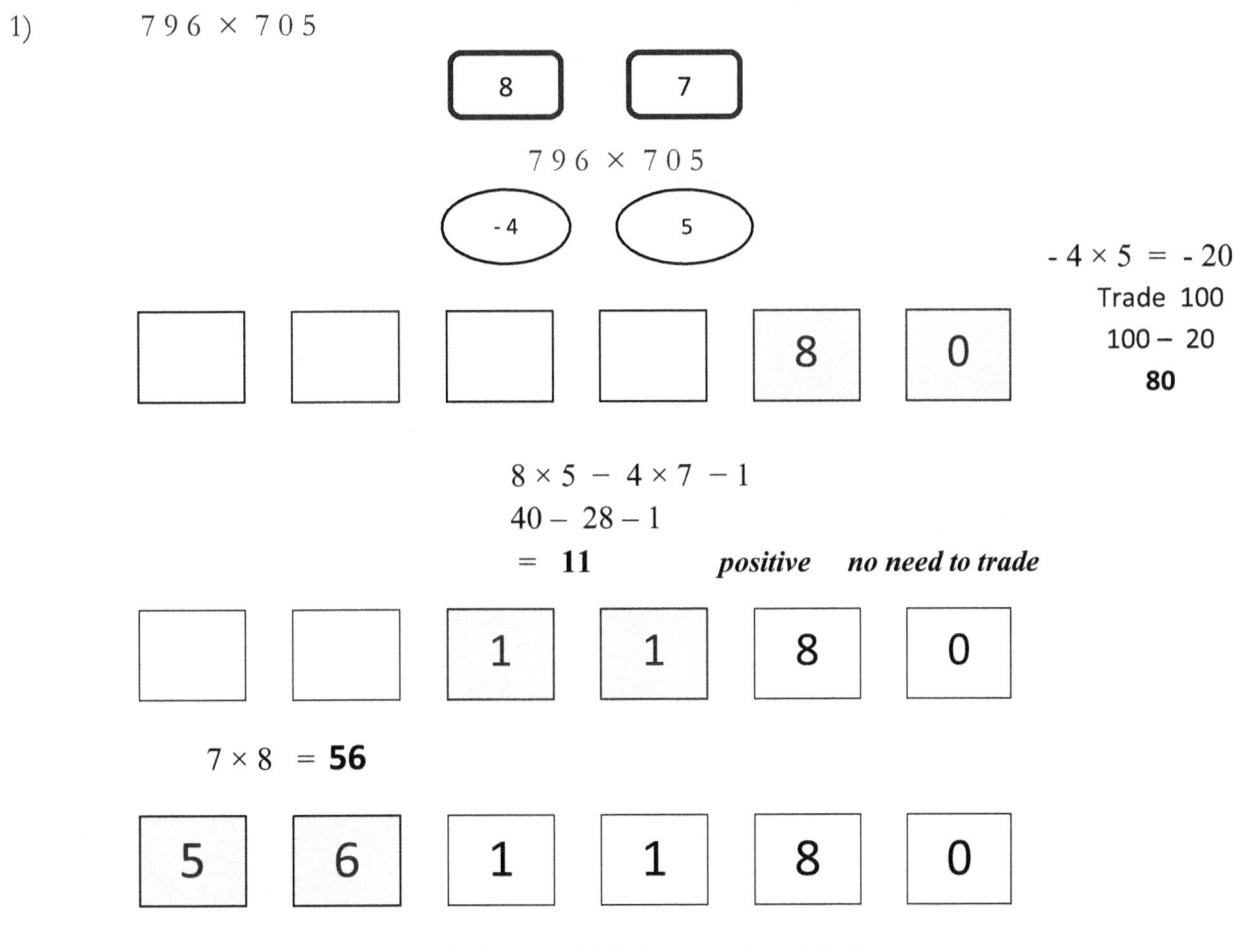

796 × 705 = 561,180

2)   492 × 309

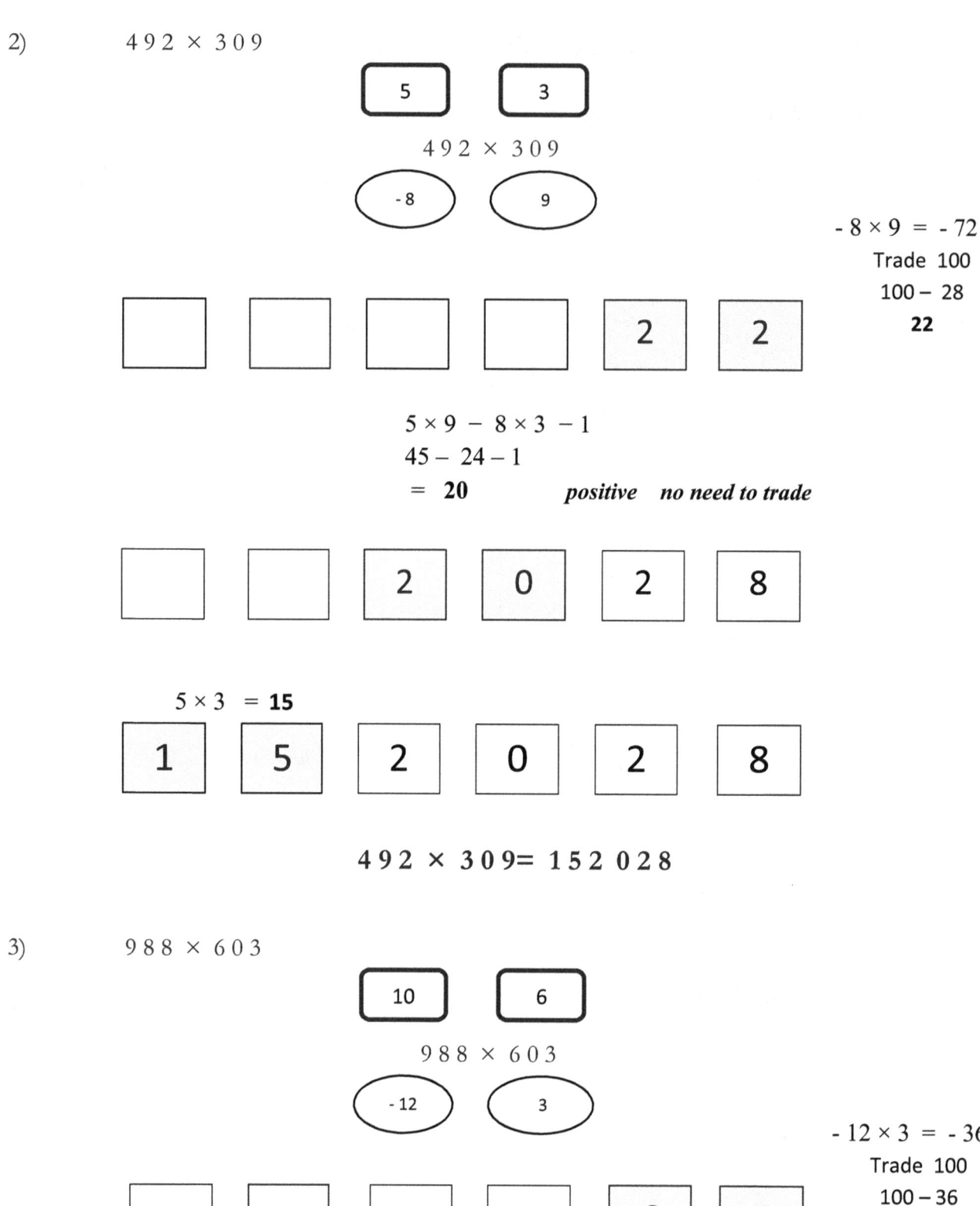

492 × 309 = 152 028

3)   988 × 603

$10 \times 3 - 12 \times 6 - 1$
$30 - 72 - 1$
$= -42 - 1$
$= -43$ *negative - trade required*

$100 - 43$
57

| | | 5 | 7 | 6 | 4 |

$10 \times 6 - 1 = 60 - 1$
59

| 5 | 9 | 5 | 7 | 6 | 4 |

$988 \times 603 = 595\,764$

4) 485×720

5   7

485×720

-15   20

$-15 \times 20 = -300$
Trade 300
$300 - 300$
00

| | | | | 0 | 0 |

$5 \times 20 - 15 \times 7 - 3$
$100 - 105 - 3$
$= -5 - 3$
$= -8$ *negative - trade required*

$100 - 8$
92

| | | 9 | 2 | 0 | 0 |

$5 \times 7 - 1 = 35 - 1$

34

| 3 | 4 | 9 | 2 | 0 | 0 |

$485 \times 720 = 349,200$

Practise questions

1) 494×608 2) 589×807

3) 492×708 4) 391×902

5) 588×711 6) 281×419

Answers

1) 300, 352   2) 475, 323

3) 348, 336   4) 352, 682

5) 418, 068   6) 117, 739

3 digit by 2 digit

1) 324 × 48

The process to solve this problem is the same for the 3 by 3 digit calculation but we use a zero in the 2nd hundreds place.

Units

4 × 8 = **2 trade (3)**

$$324 \times$$
$$048$$
$$\overline{{}^3 2}$$

Units

Tens

2 × 8 + 4 × 4 + carry = 32 + 3

5 & carry (3)

$$324 \times$$
$$048$$
$$\overline{{}^3 5 {}^3 2}$$

Tens

Hundreds

3 × 8 + 2 × 4 + carry = 32 + 3

5 & carry (3)

$$324 \times$$
$$048$$
$$\overline{{}^3 5 {}^3 5 {}^3 2}$$

Hundreds

Thousands

3 × 4 + carry = **12 + 3**

15

$$324 \times$$
$$048$$
$$\overline{15 \; {}^3 5 {}^3 5 {}^3 2}$$

Thousands

Ten thousands

3 × 0 = **0**

$$324 \times$$
$$048$$
$$\overline{15 \; {}^3 5 {}^3 5 {}^3 2}$$

Ten Thousands

$$324 \times 48 = 15\,552$$

In the following example there is no need to write the zero or do the ten thousands calculation.

2)   6 7 3 × 8 2

Units

$3 \times 2 = 6$

$$\begin{array}{r} 6\ 7\ 3\ \times \\ 8\ 2 \\ \hline 6 \end{array}$$

Units

Tens

$7 \times 2 + 8 \times 3 = 38$

8 & carry (3)

$$\begin{array}{r} 6\ 7\ 3\ \times \\ 8\ 2 \\ \hline ^3 8\ 6 \end{array}$$

Tens

Hundreds

$6 \times 2 + 7 \times 8 + \text{carry} = 68 + 3$

1 & carry (7)

$$\begin{array}{r} 6\ 7\ 3\ \times \\ 8\ 2 \\ \hline ^7 1\ ^3 8\ 6 \end{array}$$

Hundreds

Thousands

$6 \times 8 + \text{carry} = 48 + 7$

5 5

$$\begin{array}{r} 6\ 7\ 3\ \times \\ 8\ 2 \\ \hline 5\ 5\ ^7 1\ ^3 8\ 6 \end{array}$$

Thousands

$$6 7 3 \times 8 2 = 5 5{,}1 6 8$$

As you develop your skill, the working out processes will happen without the need to write down the working. But what happens if we make a mistake, as we all do? How can we know if there is an error or that our calculation is correct?

The next chapter will show you how.

"In the corporate world they pay you big bucks for thinking outside of the box!"

Chapter 7

Digit sum: Checking by 9s

If you recall in an earlier chapter on the multiplication of 9, the digits of any multiple of 9 will always sum to a single digit of 9. Thus any number of 2 or more digits, if the digits sum to a single digit of 9, then that number is a multiple of 9 or is divisible by 9.

E.g.   216   ⇔   digit sum $(2 + 1 + 6) = 9$,   therefore 216 is divisible by 9

            432   ⇔   digit sum $= 9$,   therefore 432 is a multiple of 9

But   304   ⇔   digit sum $= 7$,   therefore 304 is NOT divisible by 9

This property to find the digit sum is used to find a check digit.

When we divide a number by 9, the remainder will always equal the sum of the digits.

$10 \div 9 = 1$ remainder $\underline{1}$ sum the digits of 10 -:- $1 + 0 = \underline{1}$

$32 \div 9 = 3$ remainder $\underline{5}$ sum the digits of 32 -:- $3 + 2 = \underline{5}$

Thus the check digit is the same as dividing a number by 9. We can use this property to see if a calculation has an error. It does not guarantee that an operation is correct because the check digit is the same no matter what order of digits is in the number.

The **check digit can be used** in all four operations.

Finding a check digit can be simplified by using this fact.

If we add 9 to any number, the check digit does not change.

$$5 + 9 = 14, \quad \text{digit sum } (1 + 4) = 5$$

We can use this piece of information to simplify calculations of the digit sum. If there is a 9 or two or more digits sum to 9, we place a dot under the digit(s) and sum the remaining digits.

Sum the digits of the following numbers to find the digit sum.

    i)   345   (4 & 5 sum to 9) the digit sum ⇔ 3

    ii)   6 293   (6 & 3 sum to 9) the digit sum ⇔ 2

    iii)   25 823   (2, 2 & 5 sum to 9) the digit sum is 8 + 3 = 11 ⇔ 2

Practise questions

1)   1 4 8   2)   6 7 2   3)   7 9 3

digit sum = 4   digit sum =   digit sum =

4)   4, 6 3 6   5)   9, 5 6 7   6)   3, 2 4 8

digit sum =   digit sum =   digit sum =

7)   3 6, 4 5 2   8)   1 6, 5 8 4   9)   5 1, 9 6 8

digit sum =   digit sum =   digit sum =

Answers

2) 6   3) 1   4) 1   5) 9

6) 8   7) 2   8) 6   9) 2

A system of checking using the digit sum

When we have any operation (+, −, × or ÷), the digit sum, can help you quickly judge whether your answer has an error or is most likely correct.

Note, if the check digits are equal, it does not guarantee that the answer is correct, but you are a lot closer to making sure you have the correct answer. However, the beauty of using check digits is, that if they are not equal then you have definitely made a mistake.

When using a check digit, I will draw a circle around the number, to identify it as a check digit, ensuring that I do not use it in my calculations.

Let's look at this checking method using addition

$$34 + 14 = 48$$

Taking the digit sum of each number

$$34 + 14 = 48$$

The check digits are equal

Adding the check digits 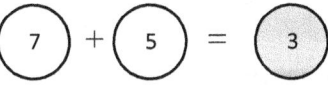 *Remember 7 + 5 = 12, then 1 + 2 = 3*

2)
$$524 + 678 = 1,202$$

Taking the digit sum of each number

$$524 + 678 = 1,202$$

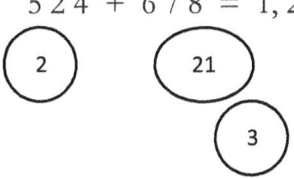

The check digits are equal

The check digits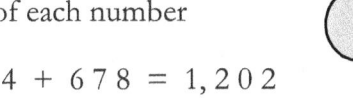

If a mistake is made in the addition

$$435 + 632 = 1,057$$

Taking the digit sum of each number

$$435 + 632 = 1{,}057$$

Adding the check digits (3) + (2) ≠ (4)

The check digits are not equal

If I have 3 or more additions, the check digits can still be used.

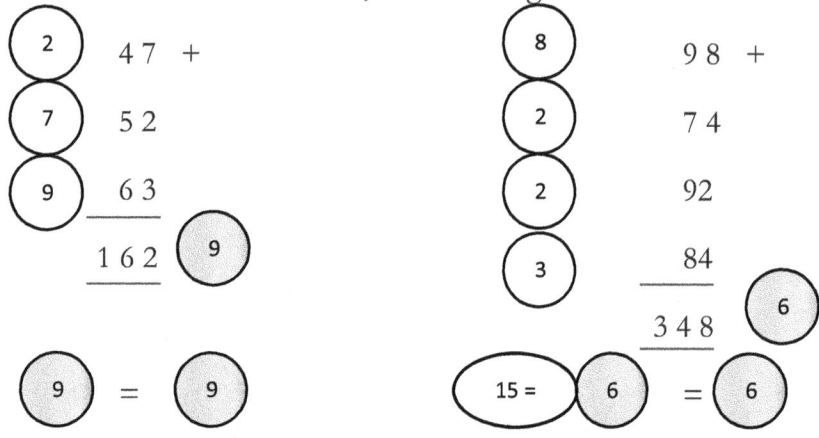

Remember we do not have to add 9s, to get the final check digit.

Thus if the check digits are equal your answer is most likely correct, but if they are not equal somewhere in your working there is an error.

Check digits using multiplication

$$43 \times 11 = 473$$

Taking the digit sum of each number

(7) (2) (5)

$$43 \times 11 = 473$$

(7) × (2) = 14 = (5)

Check digits using multiplication

(9) 45 ×
     13
   ‾‾‾‾
   = 575   18 = (9)

In multiplication, if one of the check digits is 9, then the check digit of the answer must add up to 9

⑥ 4 5 6 ×
⑥ 1 3 2
= 6 0, 1 9 2 ⑨

In multiplication if one of the check digits is 9, then the check digit of the answer must add up to 9

⑥ ⑥ × (36 =) = ⑨

Check digit method for subtraction

1)   8 3 − 2 6 = 5 7

Taking the digit sum of each number

8 3 − 2 6 = 5 7   ③

② ⑧   *At first glance the check digits do not look like they are equal*

Note the first check digit is smaller than (or equal to) the second. We can get around this in two different ways.

1. Add 9 to ② becomes ⑪ & ⑪ − ⑧ = ③

2. Rearrange the sum by adding the 2nd check digit to the check digit of the answer.

③ + ⑧ = ⑪ = ②   *Equals the check digit of the first*

You can use either method, although the first one is probably the more practical.

2)   6 4 2 − 3 6 5 = 2 7 7   ⑦
③ ⑤

Add 9 to the 1st check digit

⑫ − ⑤ = ⑦   *The check digits are equal*

3)

$$748 - 561 = 187 \quad ⑦$$
$$① \quad ③$$

Add 9 to the 1st check digit

$$⑩ - ③ = ⑦$$ *The check digits are equal*

4)

$$655 - 493 = 162 \quad ⑨$$
$$⑦ \quad ⑦$$

Add 9 to the 1st check digit

$$⑯ - ⑦ = ⑨$$ *The check digits are equal*

Check digit method for division

1)

$$36 \div 12 = 3$$
$$⑨ \div ③ = ③$$ *The check digits are equal*

However, if the 1st check digit of the dividend is smaller than the 2nd check digit of the divisor, or does not divide evenly, then as for subtraction we have to rearrange the check digits by multiplying the check digit of the quotient by the 2nd check digit.

2) *The 1st check digit is smaller than the 2nd.*

$$55 \div 11 = 5$$
$$① \div ② = ⑤$$

Rearrange the check digits for multiplication

$$⑤ \times ② = ⑩ = ①$$ *The check digits are equal*

3) $\quad 688 \div 16 = 43$

$(4) \div (7) = (7)$

Rearrange the check digits for multiplication

$(7) \times (7) = (49) = (4)$

4) $\quad 309{,}160 \div 472 = 655$

$(1) \div (4) = (7)$

Rearrange the check digits for multiplication

$(7) \times (4) = (28) = (1)$

When there is a remainder, the check digit of the **'product of digits'** is added to the check digit of the **remainder** to equal the check digit of the **dividend**.

5) $\quad 776 \div 19 = 40 \text{ r } 16$

$(2) \div (1) = (4) \quad (7)$

Rearrange the check digits for multiplication

$(4) \times (1) + (7) = (11) = (2)$

Using the method of checking by 9s to see if the given answer has an error.

1) $54 \times 9 = 486$ 2) $6,724 \times 11 = 73,954$

3) $63 \times 28 = 1,784$ 4) $767 \times 13 = 9,861$

5) $948 \times 63 = 59,724$ 6) $86,459 \times 37 = 3,198,983$

7) $657 \times 842 = 553,184$ 8) $8,136 \times 79,217 = 643,509,512$

9) $1,260 \div 84 = 15$ 10) $4,726 \div 248 = 37$

11) $5,374 \div 79 = 68 \text{ r } 2$ 12) $154,245 \div 248 = 621 \text{ r } 237$

Answers
- 1) $9 = 9$ √
- 2) $2 \neq 1$ error
- 3) $9 \neq 2$ error
- 4) $8 \neq 6$ error
- 5) $9 = 9$ √
- 6) $5 = 5$ √
- 7) $9 \neq 8$ error
- 8) $9 \neq 8$ error
- 9) $9 = 3 \times 6$ √
- 10) $1 \neq 5 \times 1$ error
- 11) $1 = 7 \times 5 + 2$; $1 = 1$ √
- 12) $3 = 5 \times 9 + 3$; $3 = 3$ √

Chapter 8

Division

"Oh No! not division!   Get me a calculator, quick!"

Ahhhhh, I remember the times when I looked at those 'eager' faces at the start of the topic on number theory. However, the look of anticipation was soon replaced with one of dread when it was time to divide a number by 2 digits. After writing **H M S bring down** on the board, I would turn around to see glazed eyes staring at me and I knew this was going to be hard work.

H M S bring down is short for

H   How many times

M   Multiply

S   Subtract

Bring down the next digit

and go through it all over again.

Not only did they have to guess how many times a 2 digit number went into the first 3 numbers, then they had to multiply that two digit number correctly, subtract the attained number from the correct digits and bring down the next digit.

Terminology

It is important to define the terms in a division operation.

$$43 \div 6 = 7 \text{ r } 1$$

- The **dividend** (43) is the number that will be divided.
- The **divisor** (6) is the number of groups that the dividend will be divided into.
- The **quotient** (7) is the number in each group that the divisor will go into the dividend.
- The **remainder** (1) is the fraction of the divisor or the number that is left over if the divisor does not go evenly into the dividend.

$$\textbf{dividend} \div \textbf{divisor} = \textbf{quotient} + \textbf{remainder}$$

The skill of division is used by students who are undertaking courses in Mathematics that involve division of polynomials (expressions with lots of powers of x).

Before we leap into division, it is highly recommend that you to practise the worksheets on single digit division, especially the exercises where you have to determine the remainder.

The method demonstrated in the following pages can be extended to 3, 4 or more digit division. But to go through that exercise requires a determination to accept a challenge of working through a new process which utilisises the techniques of the Vedic pattern of multiplication.

The first step is to go over the process when you divide a number by a single digit using the bridge method.

1)   7,456 ÷ 4

```
  4 ) 7 4 5 6
```

4 into 7 is 1, remainder 3

The remainder 3 is taken over to the next digit 4 to become 34

```
        1
  4 ) 7 ³4 5 6
```

4 into 34 is 8, remainder 2

```
        1 8
  4 ) 7 ³4 ²5 6
```

4 into 25 is 6, remainder 1

```
        1 8 6
  4 ) 7 ³4 ²5 ¹6
```

4 into 16 is 4

```
        1 8 6 4
  4 ) 7 ³4 ²5 ¹6
```

7,456 ÷ 4 = 1,864

Checking the answer   7,456 ÷ 4 = 1,864
                        ④   4   1

The check digit ④ must equal ④ × ①

                                ④ = ④

2) $65,848 \div 8$

```
          8
       _____
    8 ) 6 5 8 4 8
```

8 into 6 cannot go, therefore 8 into 65 is 8, remainder 1

```
          8
       _____
    8 ) 6 5 ¹8 4 8
```

8 into 18 is 2, remainder 2

```
          8  2
       _____
    8 ) 6 5 ¹8 ²4 8
```

8 into 24 is 3

```
          8  2  3
       _____
    8 ) 6 5 ¹8 ²4 8
```

8 into 8 is 1

```
          8  2  3  1
       _____
    8 ) 6 5 ¹8 ²4 8
```

$$65,848 \div 8 = 8,231$$

Checking the answer

$$65,848 \div 8 = 8,231$$
           (4)   (8)   (5)

The check digit (4) must equal (5) × (8) = (40) = (4)

(4) = (4)

156

3) $\qquad 634,680 \div 9$

| 9 into 6 cannot go, therefore 9 into 63 is 7 |

$$9 \overline{)6\ 3\ 4\ 6\ 8\ 0}$$

$$\begin{array}{r} 7 \\ 9 \overline{)6\ 3\ 4\ 6\ 8\ 0} \end{array}$$

| 9 into 4 is 0, remainder 4 |

$$\begin{array}{r} 7\ 0 \\ 9 \overline{)6\ 3\ 4\ {}^4 6\ 8\ 0} \end{array}$$

| You must input the 0 because it holds a place value |

| 9 into 46 is 5, remainder 1 |

$$\begin{array}{r} 7\ 0\ 5 \\ 9 \overline{)6\ 3\ 4\ {}^4 6\ {}^1 8\ 0} \end{array}$$

| 9 into 18 is 2 |

$$\begin{array}{r} 7\ 0\ 5\ 2 \\ 9 \overline{)6\ 3\ 4\ {}^4 6\ {}^1 8\ 0} \end{array}$$

| 9 into 0 is 0 |

$$\begin{array}{r} 7\ 0\ 5\ 2\ 0 \\ 9 \overline{)6\ 3\ 4\ {}^4 6\ {}^1 8\ 0} \end{array}$$

| You must input the 0 because it holds a place value |

$$634,680 \div 9 = 70,520$$

Checking the answer

$$634,680 \div 9 = 70,520$$
$$\ ⓘ○○$$

The check digit ⑨ must equal ○ × ○ = ○

⑨ = ○

157

Divisions with a remainder

4) $9,472 \div 6$

| 6 into 9 is 1, remainder 3 |

$$\begin{array}{r} 1 \\ 6\overline{)9472} \end{array}$$

| 6 into 34 is 5, remainder 4 |

$$\begin{array}{r} 15 \\ 6\overline{)9\,{}^3472} \end{array}$$

| 6 into 47 is 7, remainder 5 |

$$\begin{array}{r} 157 \\ 6\overline{)9\,{}^34\,{}^472} \end{array}$$

| 6 into 52 is 8 remainder 4 |

$$\begin{array}{r} 1578 \\ 6\overline{)9\,{}^34\,{}^47\,{}^52}\,\text{r}\,4 \end{array}$$

$$9,472 \div 6 = 1,578 \text{ r } 4$$

Checking the answer

$$9,472 \div 6 = 1,578 \text{ r } 4$$

The check digit ④ must equal ③ × ⑥ + ④

④ = ④

Practise questions

1) $972 \div 6$ 2) $927 \div 4$

3) $3,185 \div 8$ 4) $6,278 \div 9$

Answers:

1)   162   2)   231 r 3

3)   398 r 1   4)   697 r 5

2 digit Division

The technique outlined in this section is to do a 2 digit division using a single digit technique.

$$1,161 \div 43$$

Outline of the process

Our first step is to set up the bridge operation with the tens digit (4) as the divisor.

Since we are dividing by 2 digits, we draw a line before the last digit of the dividend, which separates the quotient and the remainder.

```
              quotient  | remainder
              _____
           4 ) 1  1  6 | 1
```

The units digit (**3**) is called the reduction digit and is going to be used to reduce the dividend.

```
              _____
           4 ) 1  1  6 | 1

    3
```

We perform the division as a normal single digit division.

How many times does 4 go into 11?

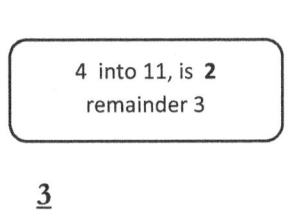

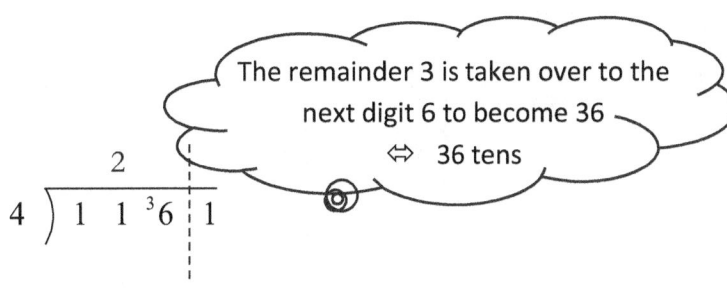

The 2 above the bridge is in the tens place, the **3** is in the units place. Therefore, its product $2 \times \underline{3} = 6$ is really 6 tens.

$$2 \times \underline{3} = 6$$

We subtract the 6 'tens' from 36 'tens' to get 30 'tens'.

```
                   2   |
              _____
           4 ) 1  1 ³6 | 1
                  - 6 |         36 − 6 = 30
    3             ____
                   30
```

With this new reduced number (30), we repeat the process:

How many times does 4 go into 30?

159

1) $1,161 \div 43$

We are going to divide by the tens digit (4).

The units digit (3) is going to be used to reduce the dividend.

We draw a dotted line separating the last digit

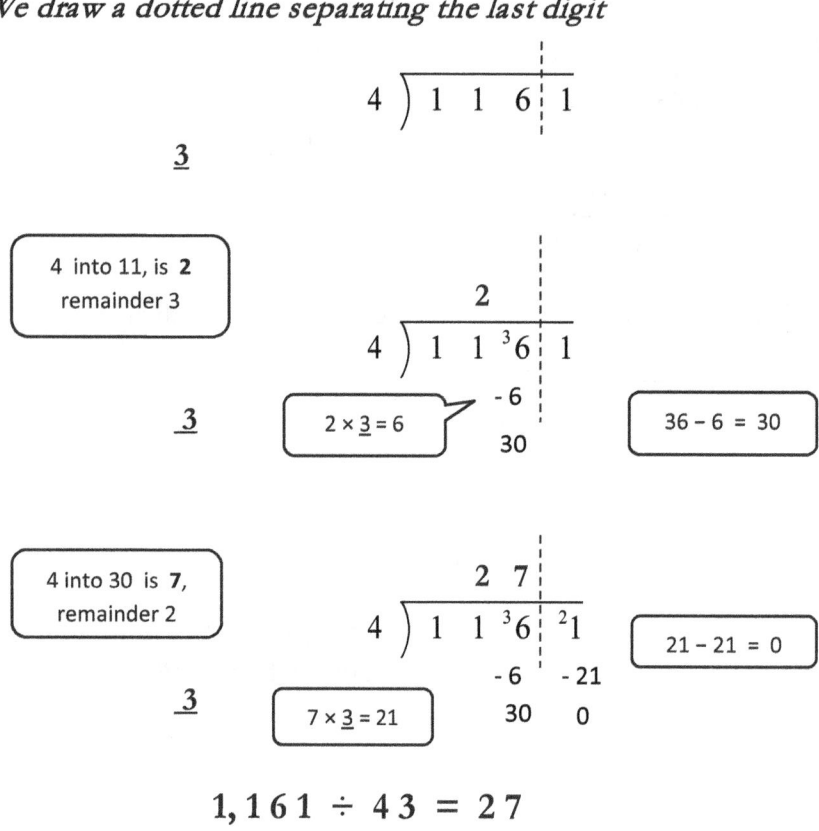

$1,161 \div 43 = 27$

Checking the answer

$1,161 \div 43 = 27$

The check digit ⑨ must equal ⑦ × ⑨ ⇔ ⑨

⑨ = ⑨

2) $4,514 \div 61$

We are going to divide by the tens digit (6).

The units digit (**1**) is going to be used to reduce the dividend.

We draw a dotted line separating the last digit

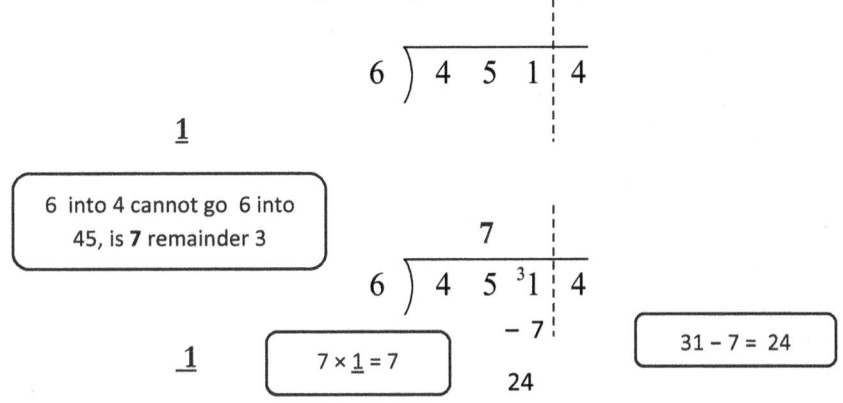

$$4,514 \div 61 = 74$$

Checking the answer   ②

$$4,514 \div 61 = 74$$

⑤ ⑦

The check digit ⑤ must equal ⑦ × ② = ⑭ ⇔ ⑤

⑤ = ⑤

3)   9,894 ÷ 34

We draw a dotted line separating the last digit

```
      ┌─────────┐
   3 )  9  8  9 ┊ 4
                ┊
   4
```

While 3 goes into 9, evenly with no remainder, the reduction factor will create a negative. Therefore, we use 3 goes into 9 twice (2) with 3 as the remainder.

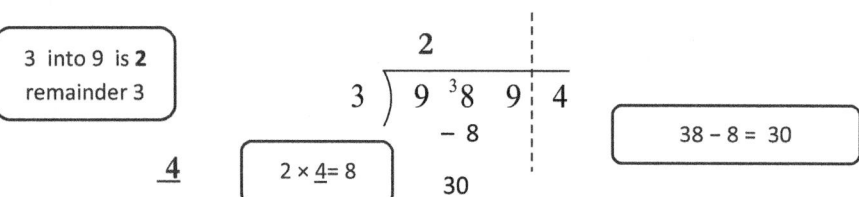

| 3 into 9 is **2** remainder 3 |

```
                2
   3 )  9  ³8  9 ┊ 4
           − 8
   4    2 × 4 = 8
           30
```
38 − 8 = 30

| 3 into 30 is **9** remainder 3 |

```
                2  9
   3 )  9  ³8  ³9 ┊ 4
           − 8 −36
   4    9 × 4 = 36
           30   3
```
39 − 36 = 3

| 3 into 3 is **1** |

```
                2  9  1
   3 )  9  ³8  ³9 ┊ 4
           − 8 −36 − 4
   4    1 × 4 = 4
           30   3   0
```
4 − 4 = 0

9,894 ÷ 34 = 291

Checking the answer

9,894 ÷ 34 = 291   (3)

(3)   (7)

The check digit (3) must equal (7) × (3) = (21) ⇔ (3)

(3) = (3)

162

4) $9,742 \div 62$

We draw a dotted line separating the last digit

```
      _____
   6 ) 9 7 4 ¦ 2
              ¦
2
```

6 into 9 is **1**, remainder 3

```
         1
      _____
   6 ) 9 ³7 4 ¦ 2
        - 2   ¦
2    1 × 2 = 2         37 − 2 = 35
        35
```

6 into 35 is **5**, remainder 5

```
         1  5
      _____
   6 ) 9 ³7 ⁵4 ¦ 2
        - 2 -10 ¦
2    5 × 2 = 10        54 − 10 = 44
        35  44
```

6 into 44 is **7**, remainder 2

```
         1  5  7 ¦
      _____
   6 ) 9 ³7 ⁵4 ¦²2
        - 2 -10 ¦-14
2    7 × 2 = 14        22 − 14 = 8
        35  44   8
```

```
         1  5  7 ¦ r 8
      _____
   6 ) 9 ³7 ⁵4 ¦²2
        - 2 -10 ¦-14
2
        35  44   8
```

$9,742 \div 62 = 157 \text{ r } 8$

While the checking process with a remainder looks difficult, it still works on the sample principles.

Checking the answer

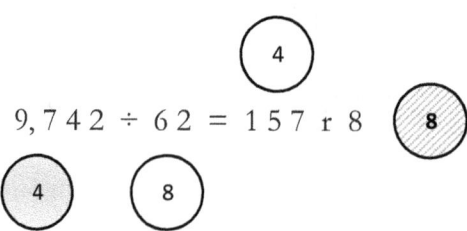

The check digit (4) must equal (8) × (4) + (8) = (40) ⇔ (4)

(4) = (4)

5) 8,394 ÷ 57

We draw a dotted line separating the last digit

```
    5 ) 8 3 9 ˌ 4
    7
```

5 into 8 is **1**, remainder 3

```
         1 5
    5 ) 8 ³3 ¹9 ˌ 4
         − 7 − 35ˌ
    7    1 × 7 = 7    26    33 − 7 = 26
```

While 5 goes into 26, 5 times with a remainder 1, the reduction number of 5 × 7 (− 35) is greater than 19. Thus we choose 5 goes into 26, 4 times with a remainder 6.

5 into 26 is 4, remainder 6

```
         1 4
    5 ) 8 ³3 ⁶9 ˌ 4
         − 7 − 28ˌ
    7    4 × 7 = 28    26   41    69 − 28 = 41
```

164

While 5 goes into 41, 8 times with a remainder 1, the subtraction of 8 × 7 is greater than 14. Thus we choose 5 goes into 41, 7 times with a remainder 6.

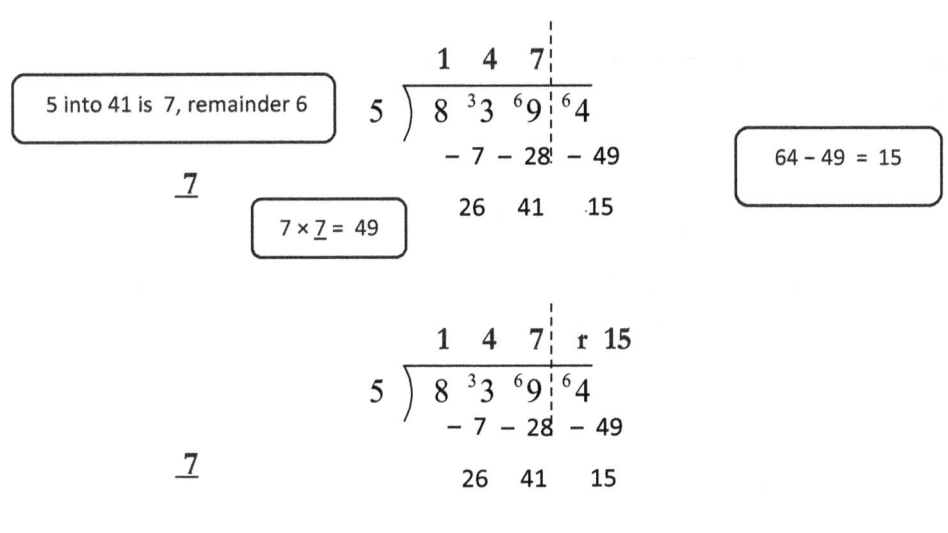

$$8,394 \div 57 = 147 \text{ r } 15$$

Checking the answer

$$8,394 \div 57 = 147 \text{ r } 15$$

The check digit ⑥ must equal ③ × ③ + ⑥ = ⑥

⑥ = ⑥

Practise questions

1) $768 \div 64$ 2) $3,921 \div 46$

Answers: 1) 12   2) 85 r 11

165

3 digit division

This section is really only for the brave hearted.

Dividing a number by three digits extends the previous process of 2 digit division. We still divide by a single digit (the hundreds digit), but use a cross multiplication technique, based on the Vedic pattern of multiplication, to reduce the dividend.

1) $\qquad 61,861 \div 432$

Our first step is to set up the bridge operation with the hundred's digit (4). As we are dividing by 3 digits, we draw a line between the third and second last digits of the dividend.

$$4 \overline{) 6\ 1\ 8\ |\ 6\ 1}$$

$$\underline{3\ 2}$$

The tens & units digit (3 2) are called the **reduction digits** and will be used in reducing the 'dividend'.

[4 into 6 is **1**, remainder 2]

$$\begin{array}{r} 1 \\ 4 \overline{) 6\ ^2 1\ 8\ |\ 6\ 1} \\ -3 \\ 18 \end{array}$$

$\underline{3\ 2}$ [1 × **3** = 3]   [21 − 3 = 18]

Thus '18' is the number that will become the 'new dividend'.

While 4 goes into 18 four times, with a carry of 2 to make '28', we still have to reduce the 'dividend' 28.

[4 into 18 is **4**, remainder 2]

$$\begin{array}{r} 1\ 4 \\ 4 \overline{) 6\ ^2 1\ ^2 8\ |\ 6\ 1} \\ -3 \\ 18 \end{array}$$

$\underline{3\ 2}$

Reducing the 'dividend' by the **expanded reduction number**.

We calculate this number by the "*outside/inside*" reduction technique.

28 will be reduced by the 'reduction number' which is calculated by

(*outside*) 4 × **3** + (*inside*) 1 × **2** = 14

```
                                1  4
  ┌─────────────┐             ─────────────
  │ 4 into 18 is 4,│         4 ) 6 ²1  ²8 │ 6  1
  │ remainder 2  │              -3  -14  │
  └─────────────┘              
         3 2                    18   14
                ┌──────────────┐                    ┌─────────────┐
                │ 4 × 3 + 1 × 2 =│                  │ 28 - 14 = 14│
                │       14      │                   └─────────────┘
                └──────────────┘
```

Why does this step work?

The **1** & **4** above the bridge are in the hundreds and tens place, and the **3** and **2** are in the tens and units place, therefore the sum of the product $4 \times \underline{3} + 1 \times \underline{2}$ is really 14 hundreds.

Thus we are subtracting 14 hundreds from 28 hundreds to get 14 hundreds.

```
                                1  4   3
  ┌─────────────┐             ─────────────
  │ 4 into 14 is 3,│         4 ) 6 ²1  ²8 │²6  1
  │ remainder 2  │              -3  -14 │-17
  └─────────────┘              
         3 2                    18   14   9
                ┌──────────────┐                    ┌─────────────┐
                │ 3 × 3 + 4 × 2 │                   │ 26 - 17 = 9 │
                │     = 17      │                   └─────────────┘
                └──────────────┘
```

Finding the remainder

Because we have moved to the right hand side of the line, we now calculate the remainder.

The digit left over from the previous reduction (9) is really 90. When we bring down the 1, it makes the amount of 91 which needs to be reduced further by 1 step.

```
                                1  4   3
                             ─────────────
                           4 ) 6 ²1  ¹8 │²6  1
                              -3  -14 │-17   ↘
         3 2                    18   14   9
                                              ↘ 91
```

This number has to be reduced further by the product of the last two digits 3 × **2**.

$$4 \overline{\smash{)}6\,^21\,^18\,^26\,1} \quad \begin{array}{c} 1\ 4\ 3 \\ \hline \end{array}$$

-3 -14 -17
18   14   9 → 91

3 × **2** = 6

3 2

91 − 6 = 85
r 85

$$61,861 \div 432 = 143 \text{ r } 85$$

Checking the answer

$$61,861 \div 432 = 143 \text{ r } 85$$

The check digit (4) must equal (9) × (8) + (4) = (13) ⇔ (4)

(4) = (4)

2)

$$84,925 \div 364$$

The hundred's digit (3), and the tens & units digit (6 4) are going to be used in the reducing of the dividend.

$$3 \overline{\smash{)}8\ 4\ 9\ 2\ 5}$$

6 4

3 into 8 is **2**, remainder 2

$$3 \overline{\smash{)}8\,^24\ 9\ 2\ 5} \quad \begin{array}{c} 2 \\ \hline \end{array}$$

− 12
  12

6 4

2 × **6** = 12

24 − 12 = 12

The '12' is the new number that will be operated on. While 3 goes into 12 four times, we still have to reduce the remaining dividend. Therefore, we try three times.

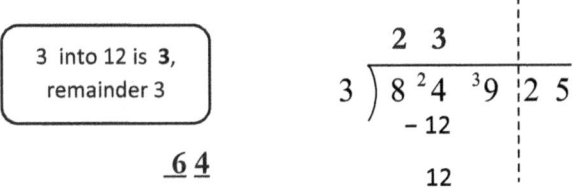

39 will be reduced by the reduction number which is calculated by

(outside) $3 \times \underline{6}$ + (inside) $2 \times \underline{4}$ = 26

[3 into 12 is **3**, remainder 3]

$$3 \overline{)8\,{}^2 4\,\,{}^3 9\,\,2\,5}$$
         − 12 − 26
6 4   12   13

[$3 \times \underline{6} + 2 \times \underline{4} = 26$]

[39 − 26 = 13]

The '13' is the new number that will be operated on. While 3 goes into 13, four times with a remainder of 1, the reduction number is too large. Therefore, we try 3 goes into 13, three times with a remainder of 4.

[3 into 13 is **3**, remainder 4]

   2 3 3
3) 8 ²4 ³9 ⁴2 5
    − 12 − 26 − 30
 6 4   12 13   12

[$3 \times \underline{6} + 3 \times \underline{4} = 30$]

[42 − 30 = 12]

Thus we are subtracting 30 tens from 42 tens to get 12 tens.

Finding the remainder

We bring down the 5, to make 125.

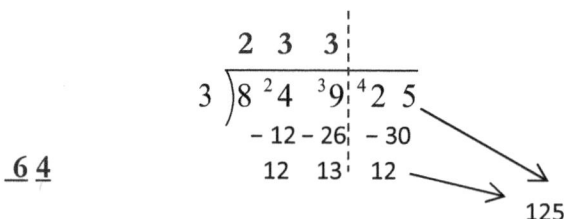

We reduce this number 125 by the product of the last two digits 3 × **4**

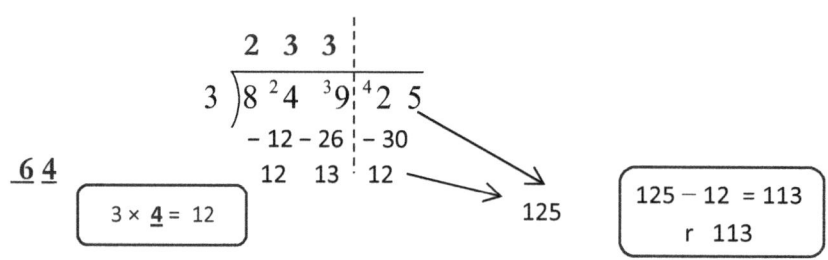

$$84,925 \div 364 = 233 \text{ r } 113$$

Checking the answer

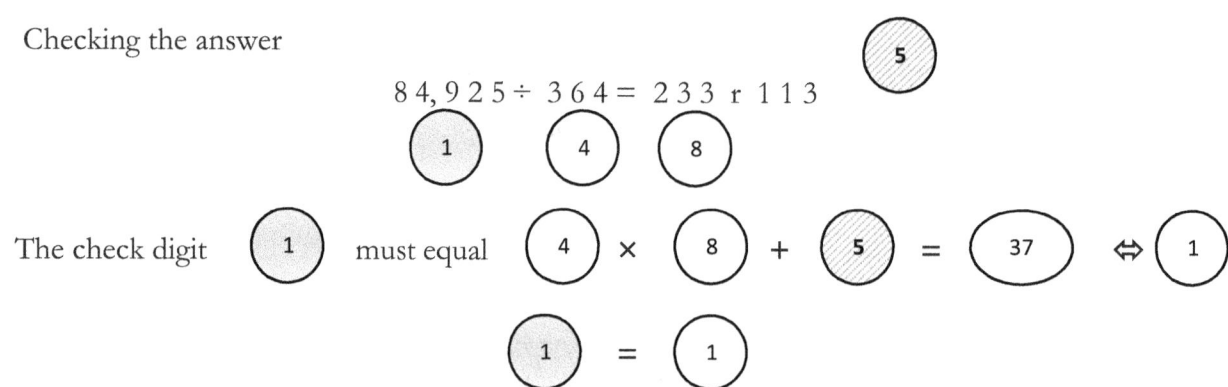

3) 74 182 ÷ 9 2 8

This example demonstrates what happens when the reduction number is greater than the 'dividend', once the initial digits of the quotient are selected. If this happens you have to adjust your digit(s) down.

The hundred's digit (9), and the tens & units digit (2 8) are going to be used in the reduction of the dividend.

$$9 \overline{)741\,|\,82}$$

2 8

9 into 74 is **8**, remainder 2

$$9 \overline{)74\,{}^21\,|\,82} \quad -16 \quad 5$$

2 8   8 × **2** = 16   21 − 16 = 5

The divisor will not go into '5' so it will be carried over.

9 into 5 is **0**, remainder 5

$$9 \overline{)74\,{}^21\,|\,{}^582} \quad -16\,|-64 \quad 5$$

2 8   58 − 64 = negative

58 will be reduced by the 'reduction number' which is calculated by

(outside) 0 × **2** + (inside) 8 × **8** = 64

Since 58 − 64 is a negative, we have to reduce the quotient's digits by 1. Instead of 80 our quotient should be 79.

9 into 74 is **7**, remainder 11

$$9 \overline{)74\,{}^{11}1\,|\,82} \quad -14 \quad 97$$

2 8   7 × **2** = 14   111 − 14 = 97

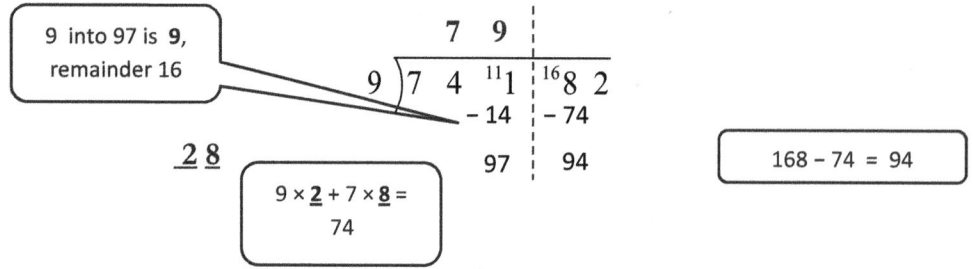

Thus we are subtracting 74 tens from 168 tens to get 94 tens.

Finding the remainder

We bring down the 2, to make 942.

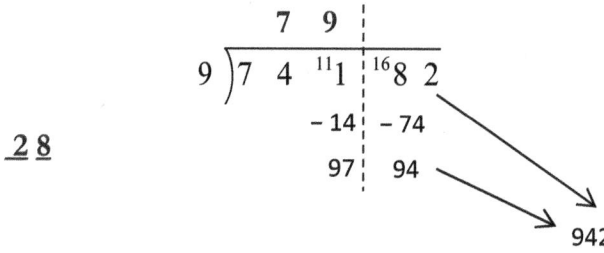

We reduce this number 942 by the product of the last two digits 9 × 8

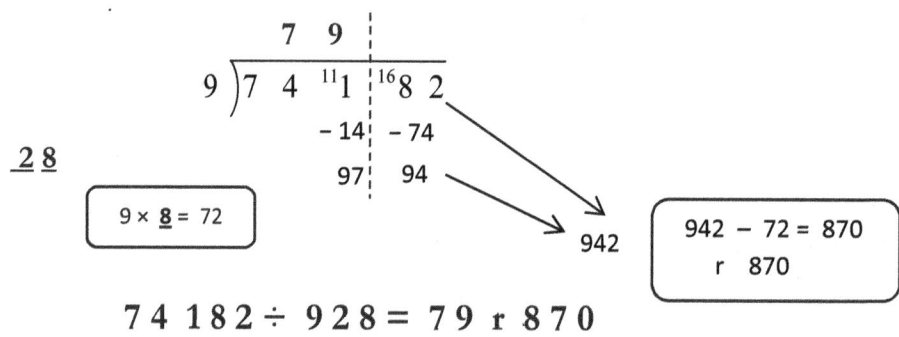

$$74\,182 \div 9\,28 = 79 \text{ r } 870$$

Checking the answer

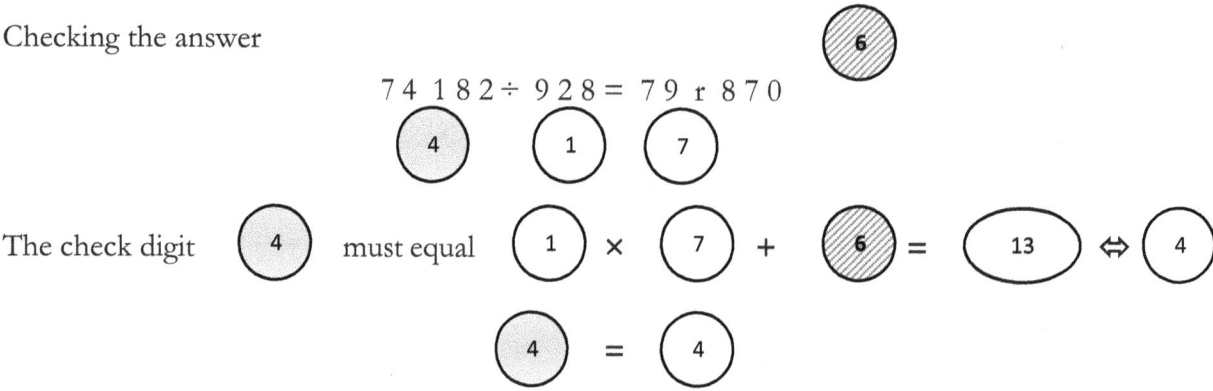

172

Practise questions

1) 9,615 ÷ 641

2) 63,921 ÷ 462

3) 88,857 ÷ 383

4) 96,278 ÷ 547

5) 8,672,839 ÷ 768

6) 3,962,195 ÷ 895

Answers:

1) 15
2) 138 r 165
3) 232 r 1
4) 176 r 6
5) 11,292 r 583
6) 4,427 r 30

Chapter 9

"In an increasingly complex world, sometimes old questions require new answers."

The Trachtenberg System

Professor Trachtenberg's system recognises that the product of two single digit numbers is in fact only a tens-digit and a units-digit. The largest calculation is 9×9 -:- 8 tens and 1 unit, the smallest excluding 0, is 1×1 -:- 0 tens and 1 unit.

Thus when we add two numbers together, no matter how large each number is, we are simply adding two digits together that have the same place value.

If we do the mental calculation, $7 \times 7 + 8 \times 9$, and treat the products as whole numbers, we may mentally say the words when we add them together.

For example: calculate the sum of $7 \times 7 + 8 \times 9$
$$= \quad 49 \quad + \quad 72$$

Summing forty-nine plus seventy-two, one approach to add these two numbers might be

> forty plus seventy is one hundred & ten,
>> plus nine is one hundred & nineteen
>>> plus two equals one hundred & twenty-one (121).

174

In calculating $7 \times 7 + 8 \times 9$,   Trachtenberg's system sees no need to calculate the products, but identify the units & tens of each product and add them separately.

units   nine & two is one (1) & carry,

tens   four & seven plus 1 is twelve (1 2)

answer   1 2 1

If we look at a two-digit by two-digit multiplication. {the 'tens' digit will be superscripted}

$$64 \times 36$$

		Th	H	T	U
units	6×4			^{2}	4
tens	6×60		^{3}	6	
tens	30×4		^{1}	2	
hundreds	30×60	^{1}	8		

So in effect, if we treat each number as a 'tens' and a 'units' digit. As we move along the steps of a multiplication, we are simply adding single digits together in their correct place value.

He calls this method the two-finger method. For right handed people using their left hand, the middle finger represents the units digit and the index finger represents the tens digit. We use these fingers to help keep track of where we are up to in the calculations.

Single digit calculation

We write the operation across the page.

1) $7 3 \times 6 =$

Units

$7\ 3\ \times\ 6$

8

Units of 6×3
8

The next calculation is the tens place value.

We know that 6 × 3 gives 1 tens digit then when we multiply 6 × 7 (which is really 6 × 70), the unit digit 2, is really a tens digit.

Tens

7 3 × 6
U   T
 3 8

Units of 6 × 7   &   Tens of 6 × 3
2   **1**
                2 + 1 = 3

Hundreds

7 3 × 6
T
4 3 8

Tens of 6 × 7
4

7 3 × 6 = 4 3 8

2)   9 4 × 8 =

Units

9 4 × 8
U
2

Units of 8 × 4
2

Tens

9 4 × 8
U   T
5 2

Units of 8 × 9   &   Tens of 8 × 4
2   **3**
                2 + 3 = 5

Hundreds

9 4 × 8
T
7 5 2

Tens of 8 × 9
7

9 4 × 8 = 7 5 2

3)   6 5 7 × 4 =

Units

6 5 7 × 4
U
8

Units of 4 × 7
8

Tens

657×4

28

Units of 4×5 &   Tens of 4×7
0   2

$0 + 2 = 2$

Hundreds

657×4

628

Units of 4×6 &   Tens of 4×5
4   2

$4 + 2 = 6$

Thousands

657×4

2 628

Tens of 4×6
2

$$657 \times 4 = 2,628$$

As you can see, we are progressively advancing across the multiplier, adding the units and tens digit of each pair.

Of course when we add to numbers, the result can be greater and equal to ten. If this happens then we can place a dot to represent the carry and add 1 in the next pair addition.

4) 738×9

Units

738×9

2

Units of 9×8
2

Tens

738×9

$^1 4\,2$

Units of 9×3 &   Tens of 9×8
7   7

$7 + 7 = 4$ & carry

| Hundreds |

738×9

U Y T ↑ Units of 9×7 & Tens of 9×3
  6 ¹4 2   3   2
                                   3 + 2 + 1 = 6

| Thousands |

738×9
↗ T ↑   Tens of 9×7
6 6 ¹4 2   6

$738 \times 9 = 6,642$

Practise questions

1) 452×8 2) $7,633 \times 9$

3) $82,503 \times 7$ 4) $972,138 \times 6$

Answers:
1)   3,616   2)   68,697
3)   577,521   4)   5,832,828

2 digit by 2 digit multiplication

The same basic technique can be extended to 2 digit, 3 digit or more digit multiplication. The advantage is that you are only adding single digits, rather than adding large numbers.

While it is difficult to express the dynamic steps of this process on a written page, I remind you that there is a PowerPoint module which will give a more dynamic and visual representation of the process.

Examining the process of 2 digit by 2 digit multiplication.

Using the connecting lines to indicate the units & tens of each product.

1)   4 6 × 6 8

When we put down the number, we write it under the unit digit.

Units

$$4\ 6 \times 6\ 8$$
$$\overline{8}$$

Units of 8 × 6
8

Tens

$$4\ 6 \times 6\ 8$$
$$\overline{^12\ 8}$$

Units of 8 × 4 + Tens of 8 × 6 + Units of 6 × 6
    2   4   6
2 + 4 + 6 =
2 & carry

Hundreds

$$4\ 6 \times 6\ 8$$
$$\overline{^11\ ^12\ 8}$$

Tens of 8 × 4 + Units of 6 × 4 + Tens of 6 × 6
    3   4   3
3 + 4 + 3 + 1 =
1 & carry

Thousands

$$4\ 6 \times 6\ 8$$
$$\overline{3\ ^11\ ^12\ 8}$$

Tens of 6 × 4
2
2 + 1 = 3

$$4\ 6 \times 6\ 8 = 3,128$$

Or the simplified U T { units tens } method

Units
```
        U⎤
   4 6 × 6 8
   ─────────
        8
```

Tens
```
      U T⎤
        U⎤
   4 6 × 6 8
   ─────────
       ¹2 8
```

Hundreds
```
      T⎤
      U T⎤
   4 6 × 6 8
   ─────────
     ¹1 ¹2 8
```

Thousands
```
      T⎤
   4 6 × 6 8
   ─────────
   3 ¹1 ¹2 8
```

2) $67 \times 98 =$

Units
```
        U⎤
   6 7 × 9 8
   ─────────
        6
```
Units of 8 × 7
6

Tens
```
      U T⎤
        U⎤
   6 7 × 9 8
   ─────────
       ¹6 6
```
Units of 8 × 6 + Tens of 8 × 7 + Units of 9 × 7
    8   5   3
8 + 5 + 3 =
6 & carry (1)

Hundreds
```
      T⎤
      U T⎤
   6 7 × 9 8
   ─────────
     ¹5 ¹6 6
```
Tens of 8 × 6 + Units of 9 × 6 + Tens of 9 × 7
   4   4   6
4 + 4 + 6 + 1 =
5 & carry (1)

Thousands
```
      T⎤
   6 7 × 9 8
   ─────────
   6 ¹5 ¹6 6
```
Tens of 9 × 6
5
5 + 1 =
6

$$67 \times 98 = 6{,}566$$

3) 78 × 64 =

Units

```
      U
   7 8 × 6 4
   ─────────
         2
```

Units of 4 × 8
2

Tens

```
     U T
      U
   7 8 × 6 4
   ─────────
        ¹9 2
```

Units of 4 × 7 + Tens of 4 × 8 + Units of 6 × 8
    8   **3**   **8**
8 + 3 + 8 =
9 & carry (1)

Hundreds

```
    T
    U T
   7 8 × 6 4
   ─────────
     9 ¹9 2
```

Tens of 4 × 7 + Units of 6 × 7 + Tens of 6 × 8
   2   **2**   **4**
2 + 2 + 4 + 1 = **9**

Thousands

```
    T
   7 8 × 6 4
   ─────────
   4 9 ¹9 2
```

Tens of 6 × 7
4

78 × 64 = 4,992

When we multiply larger numbers by 2 digits, we just continue the pattern of moving our 'fingers' applying the same units and tens addition.

4) 359 × 47 =

Units

```
        U
   3 5 9 × 4 7
   ───────────
             3
```

Units of 7 × 9
3

Tens

```
      U T
        U
   3 5 9 × 4 7
   ───────────
          ¹7 3
```

Units of 7 × 5 + Tens of 7 × 9 + units of 4 × 9
    5   **6**   **6**
5 + 6 + 6 =
7 & carry (1)

Hundreds

```
    U T
      U T
   3 5 9 × 4 7
   ───────────
       8 ¹7 3
```

Units of 7 × 3 i+ Tens of 7 × 5 + Units of 4 × 5 + Tens of 4 × 9
   1   **3**   **0**   **3**
1 + 3 + 0 + 3 + 1 = **8**

	T ⎤	*Tens of 7 × 3 + Units of 4 × 3 + Tens of 4 × 5*
	U T ⎤	2 2 2
Thousands	3 5 9 × 4 7	**2 + 2 + 2 = 6**
	6 8 ¹7 3	

Ten	T ⎤	*Tens of 4 × 3*
Thousands	3 5 9 × 4 7	**1**
	1 6 8 ¹7 3	

$$3\,5\,9 \times 4\,7 = 16{,}873$$

5) $6{,}3\,1\,8 \times 5\,9 =$

Units	U⎤ 6 3 1 8 × 5 9 2	*Units of 9 × 8* **2**

Tens	U T⎤ U⎤ 6 3 1 8 × 5 9 ¹6 2	*Units of 9 × 1 + Tens of 9 × 8 + Units of 5 × 8* 9 7 0 **9 + 7 + 0 =** **6 & carry (1)**

Hundreds	U T⎤ U T⎤ 6 3 1 8 × 5 9 ¹7 ¹6 2	*Units of 9 × 3 + Tens of 9 × 1 + Units of 5 × 1 + Tens of 5 × 8* 7 0 5 4 **7 + 0 + 5 + 4 + 1 =** **7 & carry (1)**

Thousands	U T⎤ U T⎤ 6 3 1 8 × 5 9 ¹2 ¹7 ¹6 2	*Units of 9 × 6 + Tens of 9 × 3 + Units of 5 × 3 + Tens of 5 × 1* 4 2 5 0 **4 + 2 + 5 + 0 + 1 =** **2 & carry (1)**

Ten Thousands	T⎤ U T⎤ 6 3 1 8 × 5 9 6 ¹2 ¹7 ¹6 2	*Tens of 9 × 6 + Units of 5 × 6 + Tens of 5 × 3* 5 0 1 **5 + 0 + 1 + 1 = 7**

Hundred Thousands	T⎤ 6 3 1 8 × 5 9 3 7 ¹2 ¹7 ¹6 2	*Tens of 5 × 6* **3**

$$6{,}3\,18 \times 5\,9 = 372{,}762$$

Practise questions

1) 59 × 84

2) 39 × 76

3) 82 × 75

4) 38 × 57

5) 46 × 69

6) 19 × 93

7) 659 × 74

8) 439 × 98

Answers:

1) 4,956   2) 2,964   3) 6,150

4) 2,166   5) 3,174   6) 1,767

7) 48,766   8) 43,022

3 digit by 3 digit Multiplication.

As you develop your skills in following the patterns of multiplication you will begin to improve your speed. As with any calculations errors will occur, but these will lessen as you practise the techniques.

1) $\quad 628 \times 593 =$

Units
$$628 \times 59\underset{4}{\overset{U}{3}}$$
U of 3 × 8
4

Tens
$$628 \times 59\overset{UT\ U}{3}$$
$${}^1 0\ 4$$
U of 3 × 2 + T of 3 × 8 + U of 9 × 8
6   2   2
6 + 2 + 2 = 0 & carry (1)

Hundreds
$$628 \times 5\overset{UT\ UT\ U}{9\ 3}$$
$${}^2 4\ {}^1 0\ 4$$
U of 3 × 6 + T of 3 × 2 + U of 9 × 2 + T of 9 × 8 + U of 5 × 8
8   0   8   7   0
8 + 0 + 8 + 7 + 0 + 1 = 4 & carry (2)

Thousands
$$628 \times 5\overset{T\ UT\ UT}{9\ 3}$$
$${}^1 2\ {}^2 4\ {}^1 0\ 4$$
T of 3 × 6 + U of 9 × 6 + T of 9 × 2 + U of 5 × 2 + T of 5 × 8
1   4   1   0   4
1 + 4 + 1 + 0 + 4 + 2 = 2 & carry (1)

Ten Thousands
$$628 \times \overset{T\ UT}{5\ 9\ 3}$$
$$7\ {}^1 2\ {}^2 4\ {}^1 0\ 4$$
T of 9 × 6 + U of 5 × 6 + T of 5 × 2
5   0   1
5 + 0 + 1 + 1 = 7

Hundred Thousands
$$628 \times \overset{T}{5}\ 9\ 3$$
$$3\ 7\ {}^1 2\ {}^2 4\ {}^1 0\ 4$$
T of 5 × 6
3

$$628 \times 593 = 372{,}404$$

2) $789 \times 876 =$

Units

$$789 \times 876$$
$$\overline{4}$$

U of 6×9
4

Tens

$$789 \times 876$$
$$\overline{^16\ 4}$$

U of 6×8 + T of 9×6 + U of 7×9
8   **5**   **3**
$8 + 5 + 3 = 6$ & *carry (1)*

Hundreds

$$789 \times 876$$
$$\overline{^2 1\ ^1 6\ 4}$$

U of 6×7 + T of 6×8 + U of 7×8 + T of 7×9 + U of 8×9
2   **4**   **6**   **6**   **2**
$2 + 4 + 6 + 6 + 2 + 1 = 1$ & *carry (2)*

Thousands

$$789 \times 876$$
$$\overline{^3 1\ ^2 1\ ^1 6\ 4}$$

T of 6×7 + U of 7×7 + T of 7×8 + U of 8×8 + T of 8×9
4   **9**   **5**   **4**   **7**
$4 + 9 + 5 + 4 + 7 + 2 = 1$ & *carry (3)*

Ten Thousands

$$789 \times 876$$
$$\overline{^1 9\ ^3 1\ ^2 1\ ^1 6\ 4}$$

T of 7×7 + U of 8×7 + T of 8×8
4   **6**   **6**
$4 + 6 + 6 + 3 = 9$ & *carry (1)*

Hundred Thousands

$$789 \times 876$$
$$\overline{6\ ^1 9\ ^3 1\ ^2 1\ ^1 6\ 4}$$

T of 8×7
5
$5 + 1 = 6$

$$789 \times 876 = 691,164$$

Practise questions

1) 657×349 2) 836×653

Answers:
1)   229,293   2)   545,908

An evaluation of the three methods

The three methods used to solve a 3 digit by 3 digit multiplication are.
1. The Vedic pattern
2. The 'two finger' Trachtenberg system of tens and units
3. The product method when both numbers are near base hundreds.

The fastest of the three methods is the product method, then the Vedic pattern, followed by the Trachtenberg system. With practise you should be able to mentally solve a 3 digit by 3 digit calculation in less than 30 seconds.

The Vedic pattern

Pros   The Vedic pattern can be applied in any 3 digit calculation and is not dependent on the size of each number. It is a visual method that students can easily remember and has a symmetry as we move from the units digits to the hundreds digits.

Cons   However, if the digits are higher than 5, the three middle calculations may involve mentally adding three larger numbers, thus incorporating a carry over 10 to the next calculation. To extend this pattern to 4 digit by 4 digit requires an understanding of how the pattern is constructed. The pattern works from outside in and has a symmetry. When you have 2 or more digits, the cross is always connected before the straight line.

The Trachtenberg system

Pros   The Trachtenberg system also follows a symmetry of calculations. In this method you do not have to add large 2 digit numbers together but a series of smaller single digit numbers.

Cons   The Trachtenberg system can easily be extended to 4 digit calculations and beyond. This method has more numerically smaller calculations as you are adding a units and tens digit of each product. As this system involves moving back and forth through the multiplier, you can lose your spot if you are not keeping track of the stages of calculations.

The Product method

Pros   Is a very quick method that has a simple connection between the digits in the answer and the pairs of digits in the calculation.

Cons   The ease of this method is limited to a certain range of numbers - plus or minus 20 from the hundreds. If both numbers are less than the hundreds, then the middle calculation requires you to find the difference of the number and its upper hundred. If a mistake is going to be made, it will often be at this step.

Conclusion

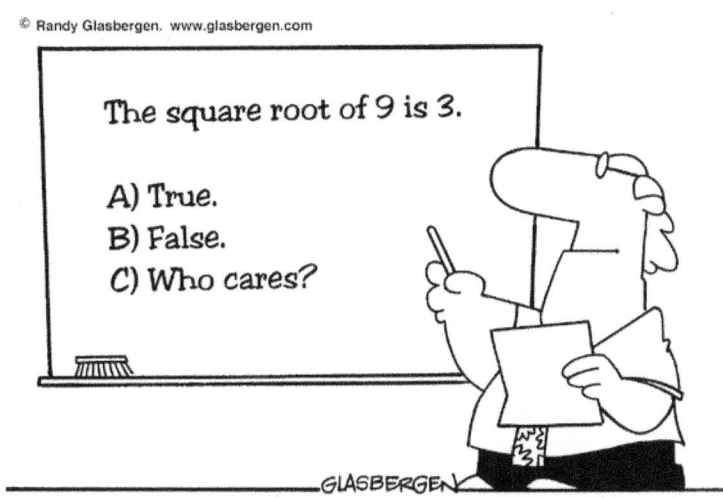

Many students actually look forward to Mr. Atwadder's math tests.

Unlike Mr. Atwadder, I do care and believe that acquiring a mastery over numbers is important.

By using the techniques in this book, it is my hope that you have gained in confidence, improved your skills and accuracy when mentally working with numbers and quickened the speed it takes to solve the four operations in Mathematics.

Numerical literacy is far more than working with numbers. It involves logical reasoning, searching for patterns, identifying trends, learning new strategies, making predictions and applying reason to solve basic or complex problems in real life situations. A numerical literate person is not afraid to face the mathematical challenges that our society presents to us, when we can easily be bombarded with statistics and numbers from Governments, Businesses and Financial institutions. By understanding the figures and arguments, people are better equipped to make informed decisions and communicate their reasoning in both oral and written forms.

I would welcome any feedback, especially if the techniques that I have demonstrated, require modification or adaption. Please feel free to contact the author at *mastering_numbers@modmaths.com.au*

Hopefully, I have been able to demonstrate that with a little understanding, that while Mathematics is not magic, it is magical.

Index Page

	page
Table of Contents	3
Introduction	5
Addition.	11
2 digit Addition	15
Subtraction	19
The trading method	20
Trading a ten for 10 units	20
Trading a hundred for 10 tens	21
Trading both a hundred and a ten.	22
Subtracting a number from a Base Number:	23
Subtraction by building up to an upper base number	27
Multiplication Terminology	34
Split and add method × 11	35
Trachtenberg's method for multiplying by 11.	36
Multiplying by 9.	41
Multiplying by 8.	47
Multiplying by 6.	51
Multiplying by 5.	56
Multiplying by 12	61
Multiplying by 7.	67
Vedic multiplication 2 digit by 2 digit	70
Shortcuts in Multiplication	77
Numbers just above 100	77
Numbers just below 100	81
Numbers on either side of 100	85
Multiplying the teen numbers.	88
Multiplying two numbers that lie between 30 & 70.	90
Two numbers above 50	92
Two numbers below 50	92
Numbers on either side of 50 -	93
Two numbers in the same decade & the units add to 10	95

	page
Two numbers above and below the same tens	97
Square numbers that end in 5	103
Square numbers close to 100	105
Squaring any number.	109
Multiplying by lots of 9s	112
The number of 9s is less than the number of digits.	115
The number of 9s is more than the number of digits.	120
Vedic multiplication of three digit numbers.	123
Both numbers are greater than the 'hundreds'	128
Both numbers less than the 'hundreds'	134
Numbers on either side of the hundreds.	139
3 digit by 2 digit multiplication	143
Digit sum: Checking by 9s	145
A system of checking using the digit sum	147
Check digits using multiplication	148
Check digit method for subtraction	149
Check digit method for division	150
Division	153
Division Terminology	154
Single digit Division	155
Single Division with a remainder	158
2 digit Division	159
3 digit division	166
The Trachtenberg System	174
Trachtenberg single digit calculation	175
Trachtenberg 2 digit by 2 digit multiplication	179
Trachtenberg 3 digit by 3 digit multiplication.	184
Conclusion	187

www.ingramcontent.com/pod-product-compliance
Lightning Source LLC
Chambersburg PA
CBHW080544170426
43195CB00016B/2671